D0152027

The Statistical Analysis of Quasi-Experiments

Christopher H. Achen

THE STATISTICAL ANALYSIS OF QUASI-EXPERIMENTS

UNIVERSITY
OF
CALIFORNIA
PRESS

Berkeley
Los Angeles
London

University of California Press
Berkeley and Los Angeles, California

University of California Press, Ltd.
London, England

Library of Congress Cataloging-in-Publication Data

Achen, Christopher H.
The statistical analysis of quasi-experiments.

Includes index.
1. Social sciences—Statistical methods.
2. Experimental design. I. Title.
HA29.A32 1986 300'.15195 85-14150
ISBN 0-520-04723-0 (alk. paper)
ISBN 0-520-04724-9 (pbk. : alk. paper)

Printed in the United States of America

1 2 3 4 5 6 7 8 9

Contents

PREFACE

Contemporary social science makes routine use of statistical methods. Economists construct models of business sectors, sociologists analyze opinion surveys, and political scientists study votes and wars. Outside academia, government agencies and businesses are experiencing a growing need for evaluation research, leading to the more frequent use of statistical techniques in decision-making. In just a few decades, quantitative approaches have come to dominate the practice of empirical social research.

Unfortunately, the new techniques are often used without much attention to the strong assumptions necessary to justify them. In public policy evaluation, for example, virtually all the methods commonly employed to estimate treatment effects assume that the subjects of the experiment have been assigned randomly to treatment and control groups. At minimum, it must be possible for the experimenter to group the subjects in such a way that, within any single group, random assignment holds. Difference-of-means tests, cross-tabulation, correlation, regression, and analysis of variance are all critically dependent on assumptions of this sort.

In practice, political, economic, or ethical considerations commonly prevent classic randomization in social science. Thus social scientists' data are often generated by forces not well described by the statistician's suppositions. In evaluation research, many a study is carried out without randomized assignment to treatment and control groups, and the actual assignment process often goes completely unspecified. Yet the same familiar statistical methods are applied to the resulting *quasi-experiment*, just as if the data had come from a randomized experiment.

Virtually all social scientists believe that useful knowledge can be drawn from nonrandomized designs. The history of their efforts to do so, however, suggests that their credo derives more from faith than from reason. Quasi-experiments have always lacked the credibility of true experiments, and rightly so. With the partial exception of economics, the social sciences have devoted little rigorous thought to nonrandomized experimentation, and most practical research methodology is supported only by intuition. Inevitably, doubts arise about the validity of particular analyses and disconcerting questions persist about the use of statistical techniques in social studies generally.

The purpose of this book is to combat these uncertainties in two central kinds of nonrandomized studies—quasi-experiments with nonrandomized treatment and control groups and quasi-experiments with nonrandom samples. The plan in each case is to describe current practice, point out its weaknesses, and suggest new procedures. Ideally, the reader will learn why contemporary analyses of quasi-experiments often yield so little reliable knowledge and discover how better methods would lead to more dependable inferences.

Chapter 1 describes the current state of data analysis in social science, with emphasis on the increasingly important field of evaluation research. Nonrandomized studies are ubiquitous, and the chapter argues that the nature of the social world makes them inevitable. Since nonrandomized experiments cannot be escaped, a rationale for them must be constructed. Social scientists have tried to do so; the success of the enterprise is currently in doubt.

Chapter 2 takes up the most common defense of nonrandomized experiments—the argument that if the appropriate variables are statistically "controlled," the analysis can be carried out just as if the study had been randomized. The chapter focuses on the special case of nonrandomized assignments to a treatment and control group, using the Coleman Report on school integration as a case study. It quickly develops that, in general, control variables are not enough. Even when certain variables affecting school achievement are controlled, as in Coleman's work, statistical biases remain. Black students in white schools will appear to perform better than their segregated peers, whether or not integration affects achievement.

Chapter 3 sets out a variety of statistical techniques for coping with nonrandomized assignments. The discussion begins with two-stage least squares (2SLS), a method by now well known to social scientists. This technique is meant to handle continuous variables in linear equations. For many purposes, however, 2SLS is inadequate. When outcomes are discrete ("success" or "failure") or when subjects can fall into just two experimental groups ("treatment" and "control"), for instance, other methods are required. The chapter sets out new methods for this situation (extensions of the linear probability model), and demonstrates that within one common class of statistical procedures it is not possible to do better. Specifically, the chapter shows how to correct for heteroskedasticity in a simultaneous equation and explains that the corresponding estimator is asymptotically best in the class of single-equation instrumental variable estimators. A similar result is given for certain nonlinear equations. A practical example from a study of pretrial release is included to show how large the biases of traditional methods can become and how conclusions change when one uses the more appropriate methods presented in this chapter. The appendix derives asymptotically efficient single-equation instrumental variable estimators for simultaneous equations that may be both heteroskedastic and nonlinear in the variables.

Chapter 4 once again examines the claim that controlling for variables eliminates bias. This time the context is the *censored sample*, a nonrandom data set drawn from the population of interest. The aim is to learn from this nonrandom sample about the cases not observed. Evaluations of college admission rules provide an instance, since they exemplify the use of a small, unrepresentative sample (admittees) to gain knowledge about the full population (applicants). That is, students in residence are studied to estimate how other applicants would have performed had they been admitted. The chapter shows again that, apart from exceptional circumstances, controlling for influential variables (proper specification) is not enough. Substantial biases may remain. These biases are derived formally in a more general setting than previously considered in the literature, and some strikingly counterintuitive consequences are discussed.

Chapter 5 reviews three different sorts of censored samples and gives estimators for each, two of them previously proposed and one suggested here for the first time. This new estimator is applied to the evaluation of a pretrial release system. In this case, the histories of the nonrandom group that was released must be used to infer the probable behavior of the jailed group had they been released. The new estimator deeply affects the analysis. A pretrial release system which seemed to perform poorly when evaluated with conventional methods is shown to be actually quite successful.

Lastly, Chapter 6 extends the methods of earlier chapters and gives advice on data analysis and model specification. Methods for handling reciprocal causation with heteroskedasticity, interaction effects, probit specifications, other nonlinearities, and time series data are explained as extensions of techniques already learned. To make the statistical work more robust, a variety of informal procedures for exploring, testing, and revising models are also given. Other topics beyond the scope of the book or for which answers are unknown are noted as suggestions for further reading and research.

Although the book contains several new econometric results, it is not written primarily for specialists in social science statistical methods. Instead it is intended for students and applied researchers whose statistical training may have stopped with a good introduction to multiple regression. The explanatory material is pitched at that level. Although some topics are more easily explained than others, most of the ideas and techniques are meant to be generally accessible to social researchers. Thus theoretical findings are stated informally and estimators are explained in ordinary language. Attention is focused on those estimators that a nonspecialist can compute dependably and easily with widely available software.

In consequence, many topics are omitted. In addition, some theoretically attractive estimators, such as full-system maximum-likelihood estimation of simultaneous probit equations, are excluded for reasons of computational complexity and modest substantive importance. Few practical researchers are likely to employ them, and their results will rarely differ from those of the simpler (linear probability) methods emphasized here. In addition, even the expert will do more preliminary data analysis with a simple estimator than with a burdensome one, and experience with the data is to be preferred to theoretical purity every time. Of course, the choice need not be exclusive. When time, software, and theoretical expertise are in ample supply, the option to compute more complex estimates, perhaps after the main part of the data analysis has been completed, is always available. Maddala (1983) is an excellent intermediate-level treatment of selection bias and related problems; Hansen (1982) and White (1984) treat many of the topics of Chapter 3 with an emphasis on asymptotics.

To keep the text readable, proofs are relegated to appendixes. There the informal language and limited perspective of the main body of the text—both inevitable when statistical language is translated into English—are remedied. The appendices assume a knowledge of econometrics approximately at the level of Theil's *Principles of Econometrics* (1971).

I hope that the empirical examples discussed here will attract readers interested in pretrial release and the justice system generally. Modern social science has taken only shallow root in the field of legal studies. The law continues to be plagued by informal speculation about human nature in the guise of legal reasoning and by the simplest kinds of ideological argument, both liberal and conservative, masquerading as economic or sociological theory. In the face of the national tragedy that is the American crime problem and the national scandal that is the American prison system, idle guesswork and windy rhetoric will not suffice. The legal system presents especially acute problems of conceptualization, data collection, and analysis, which the examples in this book have not escaped. If they stand as an illustration of how much remains to be done, however, they will have served their purpose.

Many individuals and institutions made major contributions to this book. Albert Appleton introduced me to the study of pretrial release while I worked for him in the Mayor's Office in New York in the summer of 1970. Much later, a year of full-time research was supported jointly by the Law Enforcement Assistance Administration and the National Science Foundation under NSF grant SOC78-10505. Richard Dawson and Gerald Wright in the Political Science division of NSF and Joel Garner at LEAA spent many extra hours making the dual grant possible. Alan Welsh guided me through the intricacies of the D. C. Bail Agency and generously shared his data and insights. An academic quarter of released time from the Berkeley Political Science Department under the chairmanship of Chalmers Johnson gave me important free time in which to write, and Berkeley's Committee on Research provided additional research support.

From the beginning, the project was housed in Berkeley's Survey Research Center and partially supported by that organization under the directorships of J. Merrill Shanks and Percy Tannenbaum. The center is a magnificent workplace for social science research. Without the support and shared intellectual discourse it provided, my research would have been not only less successful, but less pleasant as well.

All these individuals and groups are owed a large debt of gratitude. Of course, the opinions expressed here are my own and should not be attributed to any of them.

Valuable comments, criticisms, and encouragement were received during various stages of the project from Henry Brady, Raymond Duvall, Allan Gibbard, Duncan Snidal, Steven Rosenstone, J. Merrill Shanks, W. Phillips Shively, Garth Taylor, John Zaller, Doug Rivers, and many other friends, students, and colleagues. Since I have not always followed their advice, none is responsible for my errors.

Three very capable assistants managed the computing required by the examples. David Flanders did most of the data collection and cleaning, and he assisted in the preliminary exploration of the data. His patience, skill, and efficiency alone made the construction of the data base possible. Margaret Baker, a worthy successor in every respect, began the statistical analyses. In the final stages, Eva Eagle carried out the statistical analyses reported in the book. Her care and precision met the highest standards, and her cheerfulness was an inspiration. Thanks are also owed to Marjorie Morrisette, Gayna Sanders, and especially Fanchon Lewis, whose skill at deciphering my handwriting made manuscript preparation less painful than I deserved. Dennis Chong drew the figure in Chapter 5.

The dedication acknowledges my largest debt.

1

Experiments and Quasi-Experiments

RANDOMIZED EXPERIMENTS

Empirical research is the study of factual consequences. Do military governments accelerate economic growth? Does Head Start help poor children learn? Do standardized tests predict school performance? Analysts carry out a variety of mental tasks—theorizing, forecasting, imagining—but none of these can be performed honestly without knowing something of how social causes influence social effects. This initial step is often the hardest. In most evaluations of government policy, for example, learning how much difference the program makes is the hard part. Once the effect of the program is known, explanation, ethical judgment, and new proposals often follow quite easily.

In the purest classical paradigm, estimation of a social effect begins with the identification of a target population and the drawing of a random sample from it. Each member of the sample is then assigned randomly to either the "experimental" or the "control" group. The program is administered only to the experimental sample. In all other respects, the two groups are treated in the same way—for example, members of the sample may not be informed whether they are experimentals or controls if knowing their status might contaminate the results. At the end of the program, the two groups are compared on the outcome measures, and the differences between the two groups are taken as estimates of the effects of the treatment. The randomization in assignment to experimental and control groups guarantees that the appropriate statistical assumptions will be met, so that differences in outcomes will be unbiased estimates whose sampling errors can be easily computed.

Randomized assignment to control and treatment groups is the ideal method. Nothing else offers the same statistical trustworthiness and computational simplicity. Moreover, in a pure experiment the researcher needs no deep theoretical understanding of the variables being studied. One can learn whether a cancer treatment is effective without understanding the nature of cancer, for example. And if other factors also influence outcomes (such as the patient's general state of health in the case of cancer), one need not investigate them or even be aware of them. So long as the experimental design is faithfully executed, the experimental results will give an honest estimate of the treatment effect in the population tested.

In practice, of course, experimental designs are more complex. Human beings often react strongly and unpredictably to experimentation. If they know they will be questioned about politics, they may prepare themselves or adopt more socially acceptable attitudes (Campbell and Stanley, 1963, p. 18). If their working speed is being checked, they may work faster (the Hawthorne effect). In general, experimental effects and the reactive effects of the research itself are often commingled. Experimentalists have elaborated their techniques to meet these threats to validity (Campbell and Stanley, 1963; Cook and Campbell, 1979, chap. 3; Cochran, 1983), but the fundamental principle remains unaltered: randomization guarantees comparability of experimental and control groups, so that a treatment effect may be estimated simply by comparing the outcomes in the two groups.

THE DOMINANCE OF QUASI-EXPERIMENTS

Despite the universally acknowledged power of randomization, true experiments remain relatively rare in social and medical research. Consider these three examples from recent years:

—The Head Start program was designed to give extra education to poor preschool children so that they would be better prepared for first grade. Administrators of local programs chose children for the program with little explicit policy guidance from Washington.

Their selection procedures are unknown and presumably differed from place to place (Campbell and Erlebacher, 1975). In the best-known evaluation of the program, Head Start graduates were compared with matched children who did not enroll (Cicirelli et al., 1969). Head Start was judged to be ineffective in raising student achievement.

—The Coleman Report (1966) on school desegregation contrasted black students in mostly white schools with blacks in mostly black schools, finding that the integrated children performed better than their segregated peers. The study seemed to imply that busing for integration would raise black educational achievement.

—Breast cancer victims who undergo radical mastectomy have been compared to those who receive less drastic surgical treatments. Five-year survival rates are often quite similar for the two groups, and the conclusion is drawn that the more disfiguring surgery may be unnecessary (Sullivan, 1979).

These examples could be multiplied indefinitely. Do democratic workplaces engender democratic political attitudes (Pateman, 1970)? Do private schools produce better students than public schools (Coleman, Hoffer, and Kilgore, 1982)? Are convicted black murderers more likely to be sentenced to death than white killers (Bowers, 1974; King, 1978)? Do military regimes promote modernization in poor countries (Jackman, 1976)? Do state automobile inspections reduce highway deaths (Tufte, 1974)? Do natural left-handers forced to write with their right hands develop ulcers (Hillinger, 1979)? All these questions have been studied in essentially the same way: a naturally occurring "treatment" (say, natural left-handers trained in childhood to write with their right hands) is compared to a "control group" (natural left-handers who write left-handed). The two groups are compared on the variable of interest (ulcer rates), just as if the group assignments were random. If outcomes differ, the effect (more ulcers) is attributed to the "treatment."

In sophisticated research of this kind, additional variables may be "controlled" or "held constant". Subjects in the two groups may be matched on certain causal factors and the comparisons done within subgroups. Tufte, for instance, controls for state population density in studying the effect of auto inspections. In other situations, especially when the data cover a period of time, the subjects' prior histories may be used as a control. Thus Campbell and Ross (1968) adjusted for previous accident rates in studying the effect of a crackdown on speeding in Connecticut. However, control variables leave intact the essential logic of the simpler, two-group analyses. In each case, the experimental outcomes are analyzed as though they were the result of randomized assignment to treatment and control groups.

In fact, however, none of the studies mentioned above is randomized. They are all *quasi-experiments*—that is, something other than a random sample of the relevant population with random assignment to treatment and control groups. In particular, the preceding examples are characterized by *nonrandom* assignment. Observations enter the experimental and control groups as a result of personal decisions by the subjects or their proxies. Most black children attend schools in the neighborhoods their parents have chosen, for example, and most women select a breast cancer treatment on the advice of their doctors. The resulting experimental and control groups may be very different indeed, and nothing like the comparable groups that randomization would have produced. Even if control variables are used to fill the inferential gap, the statistical logic is quite different from classical randomization methodology.

Although nonrandom assignment studies are the most common kind of quasi-experiment, noncomparable groups can occur in more subtle fashion as well. Researchers occasionally would like to know the effect of one variable on another in a population but have data from only a subset of that population (Eklund, 1960). Thus a university might be interested in using test scores at admission time to predict school achievement (Nicholson, 1970). To learn how valid the scores are as predictors, students from previous years might be examined. But past data on achievement will be available only for those students who were admitted, not for all those who applied. The result is a nonrandom sample of applicants selected precisely for their likely success in college. However one might divide this admitted group into experimental and control groups subsequently (even by randomization), the selection procedure into college guarantees that the experiment will not be a controlled trial within the population of interest—the applicants.

Another important instance of nonrandom sampling is the opinion survey that encounters a refusal rate of 30 percent, 50 percent, or more. Since refusals are not likely to be a random group of potential respondents, the resulting sample is quasi-experimental even if the survey was designed as a random sample. Similar difficulties occur when social scientists wish to use refugees, defectors, or prisoners of war to learn about the society or the army they abandoned (Berman, 1974). Quasi-experiments of this kind are characterized by *nonrandom selection* or *selection bias*. The data they contain are said to constitute a *censored sample*.

THE WEAKNESSES OF QUASI-EXPERIMENTS

Unfortunately for everyday practice in the social and medical sciences, nonrandomized studies often fail to persuade; they leave themselves open to too many alternative explanations of their findings. Campbell and Erlebacher (1975), for example, have questioned the negative evaluation of Head Start on the grounds that the program probably selected the most deprived children. If this is true, then to compare them to unenrolled (average) children is to assess whether the experimental children made up their prior disadvantage and then some. Such a test is clearly unfair to Head Start. (See also Barnow, 1972, and Bronfenbrenner, 1975.) Similarly, black children in integrated schools may have been placed there by self-selection or by ambitious parents, as Coleman himself (Feinberg, 1978) has suggested, so that comparing them to segregated children may exaggerate the educational gains to be expected from busing. And if women with more advanced cancer select the radical treatment (Sullivan, 1979), then comparing death rates across treatments will underestimate the benefits of radical surgery.

Head Start and busing were major social initiatives of the 1960s, and radical mastectomy is a widespread treatment for one of the nation's worst medical problems. All three issues have been the subject of widespread media attention and much expensive, sophisticated research. Yet no common view has emerged on the benefits of any one of them. The disagreement can be traced directly to the nonrandomized evaluation designs. Without the power to select the experimental and control groups, researchers have been unable to arrive at a consensus.

Gilbert, Light, and Mosteller (1975) demonstrate that controversy is quite common when randomization is not employed. They studied the acceptance of research findings as a function of the quality of the experimental design. Their review of thirty-four social and medical studies led them to conclude that only randomization stills controversy. Without it, even the cleverest statistical analysis meets strong resistance from other scholars, whose professional skepticism is quite natural. When the forces that determine the experimental and control groups are unknown, the imagination has full play to create alternative explanations for the data. Inventing hypotheses of this sort is enjoyable, unstrenuous labor that is rarely resisted.

Gilbert, Light, and Mosteller, in common with most statisticians, believe that dependable inferences demand randomization. However expensive or ethically unattractive to implement it may appear, they argue, randomization gives good value. The alternative of quasi-experiments subjects people to the same regimens while leaving society as ignorant as before. Gilbert, McPeek, and Mosteller (1977, pp. 149–50) put it this way:

> When we object to controlled field trials with people, we need to consider the alternatives—what society currently does. Instead of field trials, programs currently are instituted that vary substantially from one place to another, often for no very good reason. We do one thing here, another there in an unplanned way. The result is that we spend our money, often put people at risk, and learn little. This haphazard approach is not "experimenting" with people; instead, it is *fooling around with people*.

This argument gives a utilitarian justification for randomization: we are all better off if subjects in experiments agree to enter experimental and control groups by chance.

Ignorance can be expensive in financial and in human terms—especially in medicine. In the case of breast cancer, *either* the large numbers of women undergoing the radical treatment are being unnecessarily disfigured *or* the large numbers of women avoiding it are being subjected to needless risk of death. Alas, no one knows which, and losses mount. If randomization is not permitted, the heavy price will be paid over and over for many years. And, of course, breast cancer constitutes just one of many unhappy medical dilemmas of this sort.

Certainly too little hard thinking is devoted to devising clever randomization techniques that would be attractive to the social science and medical communities. Campbell and Boruch (1975, pp. 86–94) have suggested that in instances like Head Start, in which an apparently valuable commodity is to be parceled out to a few individuals among many applicants, randomization be used to make the choices, simply because it is demonstrably less arbitrary than other schemes. If some youngsters are so in need of assistance that they must be admitted, then randomization can be applied to the remaining group. At minimum, the borderline applicants can be accepted by chance, since for them acceptance or rejection is an essentially random decision in any case.

Similarly, in medical trials, at least some doctors and patients may be sufficiently ambivalent about alternative treatments that they will agree to randomization. Randomization within that group will then give a small, but perhaps enormously valuable, true experimental trial. Generalizing from this group to the population as a whole carries risks; an experiment of this kind does not constitute a classic experiment on the relevant population. Yet at least the results will be dependable *for the randomized group*, and in some cases (especially in medicine) easy generalization to other groups will be implied. In this way, even a tiny randomized experiment may be better than a large uncontrolled study. Certainly, with ingenuity, money, and hard work, the number of random trials in social and medical affairs could be expanded; more dependable inferences and faster progress would result.

THE NECESSITY OF QUASI-EXPERIMENTS

With all this said, however, the blunt fact remains that the vast majority of interesting social and medical questions defy attempts at randomization. For these problems, controlled field experiments are impossible in one sense or another.

First of all, randomization may be *physically* impossible. Some murderers have white skin and some have black. If they commit different kinds of murders, different in ways that cannot be fully measured, then the effect of race in jurors' decisions will be confounded with that of the nature of the crime. Separating the two conventionally requires randomization. But, of course, no researcher has the ability to assign killers to one race or the other. Any knowledge of the effects of race on the adjudication of murder cases, therefore, is necessarily based on quasi-experimental evidence.

Second, controlled field trials may be *socially* impossible. Some states have chosen to have automobile inspections, and some nations are governed by the military. While these decisions are not immutable, neither are they under the control of a research team. In effect, the assignment to experimental and control groups must be taken as fixed, and all hope of randomization abandoned.

Social limits to randomization also appear in quasi-experiments with censored samples. From an inferential point of view, a university should admit a random sample of applicants, a nation should generate a random set of refugees, and an army should lose a random group of soldiers as prisoners of war. In practice, they cannot or will not. Thus controlled trials are socially impossible.

Third, randomization may be *economically* or *politically* impossible. To carry out a typical large-scale randomized study, a plan must be drawn up, administrators must be hired, records must be kept, and the cooperation of subjects must be ensured, sometimes for decades. To study the effect of rural education programs on childrearing practices in poor countries might require many hours of intrusive observation over many years if the experiment were randomized. Even a well-planned, expensive field experiment may suffer from attrition of respondents during the study as people move away or simply refuse to continue participating. By contrast, a quasi-experiment is simple: it can often use existing records and trace case histories retroactively. In similar fashion, opinion surveys with low refusal rates are very costly, many times more so than those with 40 or 50 percent responses. If the extra funds are unavailable, a random sample is out of reach.

Randomization can be priced out of the market in other ways. For example, the advantages of randomization remain ill understood by practitioners of all sorts. When randomization is in fact possible, it is often rejected or overlooked. If statistically trained researchers are called in only after a research design is in place, it may be economically infeasible to scrap it and start over. Abandoning the first design may also sufficiently threaten the administrators who planned it that they reject the new plan and thereby lose the opportunity for randomization. If any analysis is to be salvaged at that point, it must come from a quasi-experimental design.

Even if researchers are on hand early enough to suggest randomization, they may not succeed in implementing it. The subjects of experiments may simply refuse to cooperate, even when from the researchers' point of view they have no reason to refuse. For example, administrators and their clients in government programs may not *know* whether a program is effective, and in that sense randomization may not be prejudicial. Similarly, control patients in hospitals who are refused the experimental treatment may be better off: promising innovations in surgery and anesthesia are successful only about half the time (Gilbert, McPeek, and Mosteller, 1977). But staffs in public policy programs and doctors with new techniques usually *believe* them effective. In their minds, assignment to the control group means inferior treatment; hence they and their clients may resist randomization. Political power and public sympathy will frequently favor them against the isolated researchers with their "esoteric" statistical theory. The result is nonrandomized assignments.

Even when sophisticated evaluators enter a project at the earliest stages and successfully insist on randomization, a study may achieve no more than quasi-experimental rigor. A randomized cloud-seeding experiment, for instance, may encounter nonrandom days on which the seeding airplane cannot fly (Neyman, 1979, p. 91), perhaps because weather conditions are unsafe. Renting a better aircraft may be too expensive. Similarly, randomized assignment of drunk drivers to alternative treatments may fail when a clerk in traffic court decides that random numbers do not assign drivers to the treatment they "deserve" (Conner, 1978, p. 132). Unanticipated breakdowns in randomization will follow. Difficulties of this kind explain why attempted randomizations are so rare—and successful ones rarer still.

Finally, and most important, randomization may be *ethically* impossible. First of all, it goes without saying that some investigations constitute sheer abuse of scientific method. Neither the decades-long withholding of antibiotics from a control group of syphilis patients in the United States nor the horrors of the Nazi "medical studies" in World War II has the slightest justification in scientific ethics, and no one of goodwill would argue otherwise. Even in benign cases, however, when no one is deliberately maltreated, respect for individual freedom of choice may make complete randomization impossible. May a severely deprived youngster be denied Head Start because a random number table passes him over? May black students be reassigned to new schools by chance? And suppose that the evidence in favor of a particular breast cancer therapy is strong but not scientifically decisive. May women with cancer, some of them near death, be randomly directed to treatments to clinch the case? Sound ethical thinking appears to give a negative answer in most such cases.

When the effect of a treatment or program is unknown, humankind in general would probably be better off if everyone consented to randomization. Controlled field experiments may well represent the "greatest good for the greatest number," so that people who agree to enter randomized experiments perform a public service. This implicitly utilitarian argument is the principal prop for the advocates of more widespread randomized trials. But the inadequacies of utilitarian ethics are well known (Smart and Williams, 1973). In particular, this theory cannot easily take account of "rights". It overlooks the obligation not to do evil even if a greater good might thereby be obtained. One may not, for example, lynch an innocent man in the belief that a murderous race riot would be prevented. A man has a right not to be lynched.

In a similar way, the sick, the educationally deprived, and the needy have a right to assistance. When doubt exists as to which treatment is best, they have a right to choose. Of course the freedom to choose, like most rights, is not absolute. When penicillin was in short supply during World War II, not everyone who could have benefited from it had a right to it. Moreover, people may elect to waive their choice in the interests of the knowledge to be gained from randomization, and a society may wish to encourage them to do so, just as it encourages other selfless acts in the service of the common good. But the right exists nonetheless. Patients and clients may not be forced to enter randomized trials or be led into them unknowingly when they are substantially at risk and resources are sufficient to satisfy their preferences. The more serious their disabilities and involvement, the stronger their claim. It is not wrong to mislead a sophomore about the purpose of an experiment in memorization when deception is necessary. But it is wrong to place cancer patients in an experimental or control group without their well-informed consent—*even if the experiment promises enormous benefits*. Intermediate cases are more difficult to judge, but the principle remains intact. It is not enough to balance the costs and benefits of the research, especially not the costs and benefits as seen by the researcher. (Compare Selltiz, Wrightsman, and Cook, 1976, chap. 7.) At some point, the right of human subjects to choose for themselves takes precedence over the (perhaps larger) benefits to be had from coercing them.

In practice, then, randomization will often be precluded for ethical reasons. Some patients or clients will refuse to participate, and the refusals will be nonrandom. The more consequential the study, the more likely that subjects will resist randomization. Purists who insist that randomization would be better for everyone may frequently be correct, but they resemble those gentle souls who insist that if everyone loved one another, the world's problems would be solved. In much social and medical research, scientific knowledge can come only from quasi-experiments, with all their inferential pitfalls.

THE ANALYSIS OF QUASI-EXPERIMENTS

If quasi-experiments are inevitable, then their dangers should be of central concern to the social sciences. Surprisingly, among social scientists only policy analysts have been much concerned with the statistical weaknesses of the experimental designs they are compelled to use. Caveats about nonexperimental inference recur endlessly in the policy analysis literature—for example, in Struening and Guttentag's *Handbook of Evaluation Research* (1975). Outside the policy community, however, no one appears to be much concerned. Other social scientists, carrying out essentially the same statistical procedures on similar data, simply ignore the problem. The case against the nonrandomized study is well known to everyone in the social science community, but as a group only public policy researchers agonize over it.

Most researchers substitute statistical sophistication for randomized designs. Multiple regression in particular has been popular as a technique for quasi-experimental analysis. Regression (like cross tabulation, correlation, or matching) tests competing hypotheses by trying out every one. Whereas randomized studies estimate just one effect and eliminate the others by guaranteeing that they influence experimental and control groups equally, quasi-experiments must consider every explanatory possibility explicitly. Since the groups are not comparable, differential impacts from a variety of sources may have occurred. A meaningful estimate of the treatment effect depends on adjusting the experimental and control outcomes for the confounding variables that have made them incomparable. In effect, regression attempts to do this by estimating the impact of every controlled variable and then testing whether a gap in outcomes remains between the two groups. If the gap remains, the experimental effect is judged to be the residual difference. Matching, correlation, and contingency tables follow the same rules.

It is this logic, following from econometric theory, that justifies everyday practice in the empirical social sciences. The need for randomization is eliminated by holding constant factors with a substantial effect on the outcome. Some variables will go uncontrolled; the causes of social phenomena are too varied to measure them all. But if the major causes are identified and their impacts controlled, then the remainder may be too varied and unsystematic in their effects to bias the outcome. In essence, they are assumed to be randomized.

More precisely, suppose that the net effect of the uncontrolled variables is uncorrelated with both the treatment and the controlled variables. That is, the "disturbance term" is uncorrelated with all the independent variables on the right-hand side of the equation. Then the regression is said to meet the *specification* condition. This postulate is the central assumption guaranteeing that regression analysis will produce statistically attractive (unbiased, consistent) estimates of treatment effects. Matching methods, correlation, and contingency tables require the same condition. When this postulate is known to be correct, statistical theory ensures that accurate findings can be derived from quasi-experiments.

This point has been made most clearly by Cain (1975), who argues forcefully that, at least in quasi-experiments with nonrandom assignment, a knowledge of the factors influencing selection to the experimental and control groups may make randomization unnecessary. If the social class of integrated black schoolchildren differs from that of their segregated peers, then one should control statistically for social class. If sicker women are differentially present in the group receiving more radical surgery for breast cancer, then one must control for the prior state of the disease. If the researcher knows the factors biasing his results and can measure them, as he often does, then simply entering them into his regression equation will eliminate their distortion of the outcomes. The same benefits are said to derive from matching members of experimental and control groups on control variables and then comparing across matched groups (Sherwood, Morris, and Sherwood, 1975).

In practice, of course, statistical controls for variables cannot be produced mechanically. Proper specification of a statistical relation is often the hardest single step in empirical work. Nonlinearities must be assessed, variables transformed, searches for good fits carried out. Campbell and Boruch (1975) give a list of additional, less familiar specification errors that can occur in evaluation research—errors of measurement, differential response to the treatment within the experimental group, and linear trends in treatment effects over time. Any of these complexities—if left untreated—is capable of severely biasing an uncontrolled experiment. Nevertheless, each of these threats to validity is a standard problem in applied statistical work, and good methods exist to correct one and all. If Cain's argument is correct, an experienced researcher with good statistical training and familiarity with the subject matter will have little difficulty extending the basic regression model to make allowance for such effects. Their strength can then be tested in the same general way one controls other differences between the experimental and control groups. Thus in the end quasi-experiments with nonrandom assignment seem to pose no special inferential difficulties.

A similar argument could be constructed for quasi-experiments with censored samples. Indeed, researchers facing nonrandom samples apply statistical controls in the same manner and with equal confidence. The inferential adequacy of the procedure is almost never questioned.

Not everyone believes that the dilemmas of nonrandom designs can be escaped so easily. Boruch (1976, p. 167), in discussing statistical adjustment methods for the nonrandom assignment case, says that "none accommodates the specification problem satisfactorily. . . . The adjustment process is imperfect, and estimates of program effect will often be biased." Campbell and Erlebacher (1975, p. 613) use even stronger language: "The more one needs the 'controls' and 'adjustments' which these statistics seem to offer, the more biased are their outcomes."

The thrust of this argument is that applicants are often selected for programs in such a way that selection is correlated with the outcome. More successful black children may be enrolled in white schools; more deprived children may be selected for Head Start. Statistical controls for demographic characteristics or for pretreatment test scores may not capture the full extent of the prior differences between treatment and control groups. As a result, regression or matching may fail to test a program properly.

These skeptics are not arguing the trivial proposition that many factors influence a social process and hence it is difficult to control for them all. Herculean attempts at universal control are superfluous. What is sufficient for unbiased inference is proper specification: choosing a set of key factors to be controlled so that uncontrolled variables are approximately random, that is, not correlated with the controlled variables or the treatment. What Boruch and Campbell and Erlebacher are claiming is that the special circumstances present in many nonrandomized assignment experiments virtually ensure the failure of the specification postulate. They claim that policy analyses frequently exhibit biases *even when a properly specified set of variables that influence selection to the experimental and control groups has been controlled*. Adding these variables to a regression will fail to reduce bias and may even *increase* it.

This claim is paradoxical, perhaps even heretical. It has long been known that, under suitable assumptions, a properly specified contingency table or regression equation will yield the right answer. At first glance, at least, none of the usual regression assumptions seems to be violated in an ordinary nonrandomized experiment. Yet Boruch and Campbell and Erlebacher produce plausible examples and simulations in which the usual techniques fail. If their arguments are correct, at least some nonrandomized experiments pose more difficult statistical problems than is commonly supposed.

What none of the skeptics provides is a general statistical framework in which their claims can be evaluated. If their argument is correct, what characterizes the quasi-experiments in which ordinary statistical methods mislead? Which regression assumption fails? Are these inadequacies of the traditional techniques confined to a few special cases or are they routinely encountered in policy analysis and the social sciences?

Succeeding chapters take up these questions. Since it turns out that the answers depend on whether the quasi-experiment is subject to nonrandom assignment or to censoring, the two classes of quasi-experiments are treated separately. The next chapter discusses the more common class: quasi-experiments with nonrandom assignment to experimental and control groups.

2

Quasi-Experiments with Nonrandom Assignment: Why Matching and Regression Fail

THE COLEMAN REPORT

The busing of school children to achieve racial balance has been the most important and most divisive policy carried out in the public schools in the last several decades. Although the legal and social scientific justifications for it have changed over the years, at least in the beginning proponents claimed that mixing children of different races and social classes would raise the educational achievement of poor and minority children. The most prominent study to reach pro-integration conclusions was the Coleman Report (1966).

Coleman's book generated intense debate, much of it methodological. (See the collection by Mosteller and Moynihan, 1972, especially Hanushek and Kain, 1972.) The caliber of the educational measurements, the representativeness of the sample, and the validity of Coleman's regression techniques all came under heavy fire. Taken together, the criticisms severely damaged the credibility of Coleman's work—and indeed the credibility of any work carried out with those data.

The Coleman study contains many intriguing research problems, all of them worthy of further study. For present purposes, however, those difficulties will be abstracted away in favor of the general question raised by all such investigations: can statistical analysis yield up the truth in a quasi-experiment? One can conceive, for example, of an idealized Coleman Report in which black and white students were sampled randomly from around the country. For each student, the racial and social class composition of the school would be noted, along with the child's achievement scores on nationally standardized tests. The social class and educational background of the student's parents might also be recorded as control variables, along with measures of the quality of the school. Individuals matched on control variables would then be compared to assess the effect of school composition on achievement. Alternatively, ordinary regression analysis could be applied. In either case, if racially balanced schools boost minority achievement, then the racial composition variable should prove to have a large, significant effect. Coleman, essentially carrying out the regression version of this procedure, found that while racial composition made little difference, social class mixture did. Having more middle-class schoolmates apparently raised the performance of black children.

As noted in the last chapter, this finding is subject to the criticism that the study was not randomized. Black children are assigned to schools not according to a random number table but on the basis of their residential location. In turn, parents often choose a home or apartment with an eye to the quality of the school and the neighborhood's racial and social mix. Thus the composition of the school a child attends is by no means random; it reflects the character of the child's parents and therefore the character of the child as well. A priori one would expect children with high achievement scores to have parents who are better educated and more interested in having their children attend good schools—the schools that tend to be located in areas with more white, middle-class children. The "experimental group" of minority children with large numbers of well-off classmates is undoubtedly disproportionately composed of advantaged students, while the "control group" of minority children in mostly black schools is less fortunate on average. Thus even if integration had no effect, black students in white schools would be expected to perform better. These prior differences between the two groups must be adjusted statistically if the study is to be fair, and Coleman attempted to do so, for example by using regression to control for the education of parents.

Do statistical controls eliminate the bias? The Coleman Report (and for the most part its critics) proceed on the implicit assumption that Cain (1975) is correct: if the central factors that influence choice of school by black parents are controlled, then the academic consequences of segregated schools will be accurately assessed by ordinary regression analysis. More generally, it is believed, if the key variables that influence assignment to the experimental and control groups in a quasi-experiment are held constant statistically, then estimates of program effects will be accurate. As noted in Chapter 1, however, it is precisely this unstated assumption in statistical policy evaluation that Boruch (1976) and Campbell and Erlebacher (1975) have questioned.

A resolution of this dispute requires a statistical model for policy evaluations with nonrandom assignments. It will be assumed here that the sample as a whole is representative of the target population, so that the biases of nonrandom sampling are absent. (The later case is discussed in Chapters 4 and 5.) Thus the model need cope only with nonrandomness in the assignments to treatment and control groups. Such a model must formulate both the behavioral process at work and also the assignment procedure. That is, an understanding of a quasi-experiment with nonrandom assignment depends not only on how subjects behave in a given environment but also on how they got there. Within that framework, a statistical procedure can be evaluated for dependability. Since contingency tables and matching techniques are just special cases of regression, all conclusions will be stated here in terms of regression analysis.

Returning to the Coleman Report, note first that the "treatment" to which children are exposed is simply the racial or economic balance of their school. If these quantities are measured by the percentage of white students and the percentage of "middle-class" students (familial incomes above the national median, say), then both variables are continuous. Thus "experimental" and "control" are not discrete categories; exposure to better-off youngsters varies in a continuous fashion from no exposure to 100 percent exposure. From a statistical point of view, this is the simplest case to analyze.

Personal characteristics of the children affect the makeup of the schools their parents choose for them. Urban and poor children have a greater chance of being sent to schools disproportionately poor, let us suppose, while suburban middle-class children are more likely to join richer student bodies. A statistical expression for this relationship will be called an *assignment equation*: it expresses the effect of certain externally given (exogenous) factors on the assignment of the child to a treatment. In the idealized Coleman problem, suppose that the child's social class influences the choice of school. Assume also that city size affects the choice by making a greater variety of schools available in urban areas, and that this effect is the same regardless of social class. Then the assignment mechanism of interest may be expressed as a simple regression equation (with capital letters denoting variable names):

$$\text{School Percentage Middle Class} = a_1 + b_1 (\text{Middle Class}) + b_2 (\text{Urban}) + u_1 \quad (1)$$

where School Percentage Middle Class gives the class composition of the child's school; Middle Class is a variable that takes on the value 1 if the child's parents have family income above the national median and 0 otherwise; and Urban takes on the value 1 if the student's residence is in the central city of a metropolitan area of at least 250,000 residents, let us say, and 0 otherwise. The intercept is a_1, the slopes are b_1 and b_2, and u_1, is a disturbance term that captures the unexplained variation in the dependent variable. This completes the specification of the assignment equation.

The relationship of primary interest connects the treatment to the outcome in the *outcome equation*. In Coleman's case, it is the School Percentage Middle Class that is expected to influence student Achievement. To keep the example simple, just one other variable is controlled—the social class of the student. Then the outcome equation may be written as

$$\text{Achievement} = a_2 + b_3 (\text{School Percentage Middle Class}) + b_4 (\text{Middle Class}) + u_2 \quad (2)$$

The variable Urban is omitted from this equation, let us suppose, because it is known that, controlling for the wealth of the student's family and that of his schoolmates, simply living in a central city has negligible correlation with achievement scores. It is important for the forthcoming statistical methods that there be at least one such variable in a quasi-experiment—a variable that is influential in the assignment process and unimportant among the correlates of behavior.

It will be assumed that equations (1) and (2) are well specified in the sense that the control variables Middle Class and Urban constitute the key differences existing a priori among the students assigned to different schools. Other variables omitted from these two equations are assumed random in their impact—that is, uncorrelated with the effects of Middle Class and Urban. The latter two variables are therefore *exogenous;* they are uncorrelated with the disturbances u_1 and u_2. By assumption, then, equations (1) and (2) give the correct causal structure; they are well specified. Any biases that emerge in their estimation are due to the estimator, not to the omission of causal variables correlated with the included exogenous variables.

It will also be assumed that all exogenous variables are measured without appreciable error. In the present example, this assumption is unproblematical, but policy research often encounters grave observational errors, as when, for example, a score on a previous achievement test is used as a controlling variable in predicting current achievement level (Werts and Linn, 1971). Mental tests are known to have substantial errors as measures of true mental ability or achievement, and of course it is the true ability that should be controlled, not the test score. Controlling for the test score will bias the regression estimates. Correcting for errors in variables is discussed in a later chapter, but for the present the existence of other possible biases is of primary interest. To avoid contaminating them with measurement errors, the assumption of perfect measurements is made. More technical assumptions are given in the appendix to this chapter.

ORDINARY REGRESSION IN QUASI-EXPERIMENTS

The naive approach to estimating the effect of School Percentage Middle Class on Achievement is to ignore the assignment equation (1) and to simply apply ordinary least squares regression (OLS) to the behavior equation (2). This method fails, however, and econometric theory explains why this is so. The two-equation system given above is a special case of a simultaneous-equation structure. Such models are characterized by reciprocal influences among the dependent or *endogenous* variables on the left-hand side of the equation that the system is meant to explain. The system presented here is triangular; that is, the first dependent variable, School Percentage Middle Class, influences the second dependent variable, Achievement, but not vice versa. It is well known (see, for example, Hanushek and Jackson, 1977, pp. 225–43) that OLS is "inconsistent" when applied to simultaneous equations in general and to triangular systems in particular. Roughly speaking, this means that apart from lucky accidents due to sampling variation, regression gives the wrong answer no matter how much data is available.

The problem can be avoided in triangular systems if the disturbance terms u_1 and u_2 are uncorrelated with each other. Triangular systems with this property are said to be *recursive*. In that case, ordinary regression gives a dependable estimate with the usual good statistical properties. However, the assumption of uncorrelated error terms strains credulity in most policy evaluations. The complete set of factors that influence assignment can never be measured, nor can all the variables that cause the outcome variable. Many if not most of these unobserved quantities affect both assignment and outcomes. In Coleman's problem, for example, parents' education, their concern for their children's future, their belief in education, their children's attitude toward school, the attitudes of the children's friends, and a great many other factors will surely influence both the child's chance of being in a more middle-class school *and* his or her achievement level. These variables can never all be measured, and the unmeasured ones will enter both u_1 and u_2. The result is a correlation between the disturbances and erroneous estimates when OLS is applied to equation (2).

The precise distortion of the regression estimates in the outcome equation of an arbitrary two-equation triangular system can be derived by the methods described in the appendix to this chapter. The discrepancy is best expressed as an *inconsistency*, which is a kind of asymptotic bias. (See the appendix.) It turns out that

$$\text{Inconsistency in treatment effect} = \frac{\sigma_{12}}{s^2} \tag{3}$$

where σ_{12} is the covariance between u_1 and u_2 and s^2 is the unexplained variance in the assignment variable when it is regressed on the exogenous variables in the outcome equation (2).

This result shows first that the asymptotic error in estimating the treatment effect is proportional to the covariance between the disturbance in the two equations, σ_{12}. In the case of the simplified Coleman equations (1) and (2), the larger the number of unmeasured variables that influence both the choice of a school and the child's performance in it, the larger σ_{12} will be. In this instance, the unmeasured variables would be expected to work in the same direction in both equations, making σ_{12} positive. Ambitious parents are likely to enroll their child in a school with better-off children *and* have children who perform above average for their circumstances and social class. Thus, as one might expect intuitively, estimating the behavior equation with ordinary regression puffs up the apparent effect of having well-off classmates, all else equal. The unmeasured effect of ambitious parents is attributed to the composition of the school, with which it is correlated. Thus regression will seem to show that middle-class schools raise the performance of children, even when control variables are introduced—whether or not a genuine effect exists. They might do so even when the actual effect is negative—if the inconsistency (Equation 3) is large enough to overwhelm the true negative effect. Thus regression may indicate precisely the opposite of the truth.

As noted above, the sole escape from this bias occurs when *no* unmeasured variables influence both the assignment process and behavior.[1] Essentially, assignment to treatment and control groups must occur purely randomly or purely mechanically. If children were assigned to schools purely by chance, for example, the random error term in the assignment equation would not correlate with unmeasured variables in the behavior equation. The numerator σ_{12} in the bias term would then necessarily be zero.

Alternatively, in a perfectly segregated neighborhood school system in which every child is assigned the nearest school of his or her race, the assignment equation is deterministic: race and neighborhood determine the school perfectly. There are *no* unmeasured variables in the assignment equation and hence none that influence both assignment and behavior. Thus σ_{12} will be zero.

With random or deterministic assignments, then, bias evaporates as long as the other assumptions are met. In particular, the assignment

---

[1]It is technically possible that several unmeasured variables might appear in both u_1 and u_2 but their net effects exactly cancel out in the correlation, making σ_{12} zero and eliminating the bias. This event is too unlikely to be of practical importance.

variables measuring race and neighborhood must be controlled in the outcome equation to ensure that they are uncorrelated with the disturbance in that equation. In these instances, nothing more need be done to guarantee a consistent estimate of the treatment effect.

Cain (1975) gave several hypothetical examples in support of his thesis that controlling for assignment variables eliminates bias in general. Inspection of these examples shows that his assignment processes are either random or deterministic. For such cases, his argument is correct. Suppose that clients for a program are chosen by a purely mechanical rule known to the researcher, by a purely random mechanism, or by some combination of the two (randomization for clients above a certain cutoff score on a pretest, for example), and everyone else is in the control group. Then simple regression applied to the behavior equation will be unbiased and consistent so long as any variables used in the assignments are controlled in the outcome equation. This is the sort of special case usually treated in books on quasi-experimentation (see Cook and Campbell, 1979, chap. 4).

Unfortunately, policy programs rarely employ such simple selection mechanisms. Ordinarily clients select themselves for a program, or administrators choose them from among a group of volunteers by processes that are neither deterministic nor purely random. Then σ_{12} will not be zero. Quasi-experimental data with nonrandom assignment are distinguished from true experimental data by the likelihood that in the quasi-experiment, $\sigma_{12} \neq 0$. The larger this covariance, all else equal, the worse the regression estimates will become.

CONTROLLING FOR VARIABLES IN A QUASI-EXPERIMENT

The second noteworthy aspect of the inconsistency (Equation 3) is its denominator, s^2. Now s^2 is the variance unexplained when the exogenous variables in the outcome equation are used to predict assignment. A good fit yields a small s^2 and thus a large inconsistency.

In the Coleman model, the sole exogenous variable in the outcome equation is Middle Class. Using the exogenous outcome variables to predict assignment to a treatment in this case means employing just Middle Class to predict the choice of School Percentage Middle Class. If Middle Class does well on its own in explaining school selection, then the inconsistency will be large.

To see intuitively why this should be so, consider the outcome equation (2). In this equation, students in middle-class schools are compared to their peers in lower-class schools. The comparison is not based on the overall difference in performance between children in the two types of schools but rather on the difference in performance for children of the same social class, since Middle Class is controlled. That is, middle-class students in mostly middle-class schools are compared to middle-class children in mostly lower-class schools. The same thing is done for lower-class children in the two kinds of schools. This procedure can result in comparisons between groups very different in ambition, however, and the ensuing estimate of the treatment effect can be even more erroneous than without the control for social class.

Suppose for a moment that being middle class has no direct influence on school achievement—that, in similar circumstances, lower and middle-class children achieve at the same level. Then the difference in performance between children in middle-class schools and those in lower-class schools would be due entirely to the social class of schoolmates and to the ambition of the children who enter the two types of school. In this case, Middle Class would be worthless as a control variable in the outcome equation (2).

The child's social class does influence the choice of school: lower-class children are less likely to enter middle-class schools. (See Equation (1).) Those who do so will be disproportionately overachievers. Simply being in a middle-class school gives evidence of familial ambition; being in a middle-class school in spite of being lower class gives evidence of even stronger ambition. In this example, then, controlling for Middle Class in the outcome equation is equivalent to comparing the achievement of lower-class children in middle-class schools (who are on average *highly* ambitious) to lower-class children in lower-class schools (who are not). This comparison considerably exaggerates the true impact of attending a middle-class school. The more powerful Middle Class is in determining school choice, the harder it will be for poor children to enter middle-class schools. And the harder it is for them to enter, the more highly ambitious will be the few lower-class students who do enroll in middle-class schools—and the more erroneous the estimate of the treatment effect will become. Thus one can see why the influence of Middle Class on school choice is important. The better Middle Class predicts the choice, the worse the inconsistency, just as Equation (3) suggests. A strong relationship between Middle Class and school choice reduces s^2, which inflates the inconsistency.

Notice that in this example a researcher would be better off not to control for social class at all in the outcome equation. Without the variable Middle Class in the equation, all children in middle-class schools (most of them middle class and not particularly ambitious) would be compared to all children in lower-class schools (most of them lower class and not particularly unambitious). The resulting two groups would differ little in average ambition, and since social class is assumed not to matter, their difference in achievement would be due primarily to the percentage of middle-class pupils in the school—that is, to the treatment effect. Controlling for the useless variable Middle Class, so that all comparisons are carried out within social classes, makes the inconsistency worse. But the researcher using OLS will be unaware of all this. Middle Class will appear to have an effect on outcomes (see the appendix), and its status as a legitimate control variable is likely to go unquestioned.

What the foregoing example demonstrates intuitively is this: in estimating a treatment effect with quasi-experimental data, controlling for variables in the outcome equation that have no real effect—but do influence assignment—will always worsen the inconsistency.[2] Moreover, the worthless control variables will appear useful. The more the variables influence assignment, the more inconsistent the estimates tend to become. This result contrasts sharply with the

---

[2]This conclusion follows somewhat more formally from the form of the inconsistency (3). When a variable such as Middle Class has no genuine effect on achievement, its coefficient is zero. The behavioral relationship is the same with or without that variable entered; in particular, the disturbance term u_2 is the same in either case, so that its correlation with u_1, namely σ_{12}, is also unaltered. Hence the numerator of the inconsistency is not affected by the inclusion of the additional controlling variable.

On the other hand, having Middle Class in the behavior equation will certainly reduce the denominator s^2, the unexplained variance in selection when it is regressed on the exogenous behavior variables. By hypothesis, Middle Class influences selection, so that having it in the behavior equation will reduce s^2. In sum, then, the numerator of the inconsistency is unchanged and the denominator is reduced, guaranteeing that the fraction will increase in absolute value. There is no offsetting gain from reduced specification error: the new variable has no genuine effect on behavior. Hence the inconsistency can only increase.

ordinary regression case, in which controlling for variables without any effect causes no inconsistencies. With enough data, one gets the right answer. In quasi-experiments, the control variables can be positively harmful no matter how large the sample.

Up to this point, it has been assumed that the outcome equation is correctly specified. In practice, of course, specifications are never known to be true, and variables are added to equations in hopes of reducing the error. In particular, if just one exogenous causal factor has been omitted from an equation, including it will always eliminate any inconsistency due to specification error. Thus adding a variable with a genuine effect to the outcome equation will improve the estimate, all else equal.

Unfortunately, *ceteris paribus* does not hold in quasi-experiments. Both the misspecification and the simultaneity of the equations contribute to the inconsistency; reducing the former typically increases the latter. This result occurs because an exogenous variable added to an outcome equation typically has an impact on assignment also. As seen in the previous example, controlling in the outcome equation for a variable that influences assignment to a treatment tends to increase the inconsistency. This increase may or may not be enough to offset the reduction in specification error; there is no necessary reason why one effect should be larger than another. Thus the pessimistic conclusions of the preceding example can extend also to the case in which the control variables have a genuine impact on outcomes and belong in the equation. *With quasi-experimental data derived from nonrandomized assignments, controlling for additional variables in a regression may worsen the estimate of the treatment effect, even when the additional variables improve the specification.*[3]

The same argument applies when contingency tables or matching procedures are used (McKinlay, 1975). Cain's (1975) advice about correcting for assignment effects by controlling for additional variables does not in general extend beyond the rather special cases he considered. In quasi-experiments, controlling for factors that influence assignments more than outcomes may well make matters worse.

---

[3]A more formal demonstration of this conclusion may be constructed along the lines of note 2.

This aspect of the OLS inconsistency in estimates of treatment effects may be regarded as a kind of multicollinearity problem, though its effects are quite different from those of ordinary multicollinearity (see, for example, Theil, 1971, pp. 179–81). In a quasi-experiment, outcomes are influenced by the treatment plus control factors. The more correlated these two sets of variables, the lower s^2 will be and the larger the inconsistency will become. The degree of correlation among the independent variables in a regression is described as their *multicollinearity*. When the treatment variable is highly collinear with the control variables in a quasi-experiment, this form of multicollinearity increases the inconsistency of the regression. Adding control variables to a policy outcome regression is most dangerous when these additional factors increase the multicollinearity substantially without materially improving the specification.

A CLASSIC TECHNIQUE FOR REDUCING INCONSISTENCY

Considerable ingenuity has been expended in devising methods to cope with the evident regression inconsistencies that arise in quasi-experiments. (see, for example, Cochran, 1968; Cochran, 1983; Nunnally, 1975; Werts and Linn, 1970, 1971). The most popular of these methods attempts to "partial out" the effect of the exogenous variables in the behavior equation before the treatment effect is estimated. In the Coleman equations (1) and (2), Achievement might be regressed initially on Middle Class to remove the effect of social class on performance. Then the residuals from this regression are taken to represent achievement above the level expected for that class. They may then be regressed on School Percentage Middle Class to assess the effect of social class composition on achievement. A two-step technique of this kind is used so commonly in the study of test score changes that Werts and Linn (1971, p. 17) regard it as "the traditional procedure."[4]

This method fails, too. Goldberger (1961) has pointed out that in an ordinary regression problem, two-step regression methods of this

---

[4]Note that two-step regression is quite different from two-stage least squares. The latter technique is discussed in Chapter 3.

kind are biased. He notes in particular that when just one coefficient is estimated at the second step (the treatment effect), the effect is underestimated (that is, reduced in absolute value). The expected effect is reduced to $b_3(1 - R_{21}^2)$, where b_3 is the true effect and R_{21} is the multiple correlation between the control variables entered at the first step and the treatment variable. Thus even when none of the special statistical difficulties arising in quasi-experiments are present, this technique will mislead. (See also Goldberger and Jochems, 1961.)

Goldberger's result can be extended easily to the typical policy evaluation problem with nonrandomly assigned treatment and control groups. As we have seen, ordinary regression fails to produce the correct estimate of the treatment effect in a quasi-experiment. The two-step estimator reduces the size of this erroneous estimate by the fraction R_{21}^2. But there is no reason to think that this reduction will exactly cancel the inconsistency or even bring the estimate nearer the truth. If the regression estimate is too small, for example, then the two-step method will reduce it still further, making the inconsistency worse. In general, then, the two-step procedure is no more helpful than ordinary regression.

A modified version of this procedure (suggested by Cain, 1975, pp. 314–15) is less harmful, though it is applicable only in special circumstances. Suppose that the researcher has a genuine control group selected randomly from the population. This group is exposed to no treatment, so that its performance depends only on exogenous variables. Then regression of outcomes on these factors in the control group sample will give honest estimates. Now if the effect of these variables is the same for the treatment and control groups, these same coefficients can be used to estimate what the treatment group's performance would have been in the absence of the treatment. The residuals from these predictions can then be regressed on the treatment variable across the full sample to estimate its effect.

This procedure is better than applying ordinary least squares to the entire sample, but it is inconsistent all the same. An analysis like that given in the appendix shows that the inconsistency formula used throughout this chapter, namely σ_{12}/s^2, is appropriate once again. Some reduction in inconsistency occurs because s^2 is the unexplained variance in treatment assignments when the exogenous outcome variables are used as regressors. With the Cain techniques, there are no exogenous variables left in the outcome equation; their effect has been removed at the first step. Hence the unexplained variance s^2 is the total variance in the treatments. Thus s^2 is maximized, which reduces the inconsistency. As long as some unmeasured variables influence both selection and outcomes, however, so that $\sigma_{12} \neq 0$, inconsistency will remain. This procedure, too, fails to estimate treatment effects accurately.

SUMMARY

The basic conclusion to be drawn from this analysis is that neither simple regression nor any of the adjusted regression techniques discussed in this chapter will eliminate erroneous estimates. In a practical problem, where important factors that influence assignment and behavior typically go unmeasured, the estimated treatment effect will generally not converge to the true effect, no matter how much data is obtained. In certain cases, moreover, even the estimated *direction* of the effect may be reversed from the truth. The appendix shows that this result is quite general: it matters little whether the treatment is a continuous variable, such as School Percentage Social Class, or a dichotomous variable, as in a treatment-and-control-group design. Nor does it matter whether the behavior variable is continuous, as in the case of school achievement scores, or dichotomous, with categories like "success" and "failure." The treatment effect will be estimated incorrectly in all such cases. In fact, the apparent effects of all the variables in the behavior equation will be erroneous. Factors with no real impact may appear to be influential; important variables may have vanishingly small coefficients. Specifications that seem to improve the fit may lead the researcher further from the truth. In general, no statistics from the behavior equation can be trusted.

For the reader with some background in simultaneous-equation estimation, none of these findings will be surprising. When people can *choose* their scores on one of the causal variables in a social situation, ordinary regression will fail. If the type of treatment can be selected by the subjects (or by someone like them, as when their parents enroll them in a school, or by someone who takes their characteristics into account, as when a doctor chooses a therapy for a patient), then the treatment is no longer exogenous, no longer given. A nonrandom experiment is a set of linked stochastic processes, and the elementary techniques usually applied in policy analysis will not disentangle the causal mechanisms at work in each of them.

Policy evaluations and other social science studies are likely to be carried out by statistically inadequate methods for some years to come. If one knows the formula for the inconsistency, however, and has some intuitions about the probable magnitude of the quantities in it, as frequently happens, a reasonable guess about the truth can be made. In problems like those considered in this chapter—when members of the treatment and control groups together constitute a random sample of the population but assignment between them is carried out nonrandomly—the inconsistency is just σ_{12}/s^2.

One must especially beware of two situations. First, important variables that influence both assignment and behavior are omitted from the outcome equation. Then σ_{12} will be large and the effect of the treatment will be seriously misjudged. The direction of the inconsistency depends on the omitted factors: if the same factors that cause entry into the treatment group also raise the level of the outcome measure, then the two effects are positively correlated and the treatment effect will be exaggerated upward. If entry into the control group is correlated with lower outcome scores, the treatment effect will be biased downward.

In the second case, substantial multicollinearity exists among the causal variables in the outcome equation. This situation is dangerous because the size of the inconsistency is affected by s^2, which, as noted earlier, can be regarded as a measure of the collinearity among the variables influencing outcomes. When multicollinearity is high, in the sense that the causal variables other than treatment can be used to predict treatment well, then s^2 will be reduced and the inconsistency will rise. As we have seen, this effect can be so large that one is better off omitting some of the true causal variables.

The frailty of ordinary regression when confronted by simultaneous statistical equations has been recognized for nearly forty years (Haavelmo, 1947), and straightforward computational techniques for the simplest cases have been available for more than two decades (Basmann, 1957; Theil, 1958). The need for new methods in quasi-experimentation has been pointed out in informal and accessible articles for more than ten years (see, for example, Dykstra, 1971). Yet the appropriate methods are virtually never applied in practical policy evaluations, and their use in social science remains rare outside economics.

The intellectual lag stems primarily from simple ignorance of econometric techniques. But another cause is the nature of policy evaluation and social data. The discontinuous or discrete variables employed so commonly outside economics raise special problems in simultaneous-equation estimation, problems that have only recently begun to be addressed. The next chapter reviews the state of the art in the estimation of reciprocal causal systems, including those with discrete variables, and goes on to suggest manageable methods for the analysis of nonrandom assignment quasi-experiments and related problems.

Appendix

Consider first the case in which the treatment is a continuous variable. The selection equation can then be written as

$$y_1 = X_1\beta_1 + u_1 \tag{A-1}$$

where y_1 is an $n \times 1$ vector of observations on the treatment, X_1 is an $n \times k_1$ matrix of observations on exogenous factors influencing selection to the treatment, β_1 is a $k_1 \times 1$ vector of coefficients to be estimated, and u_1 is an $n \times 1$ disturbance term.

The outcome equation is specified as

$$y_2 = \gamma y_1 + X_2\beta_2 + u_2 \tag{A-2}$$

where y_2 is an $n \times 1$ vector of observations on the outcome of the experiment, X_2 is an $n \times k_2$ matrix of exogenous variables, γ and β_2 are 1×1 and $k_2 \times 1$ dimensional coefficient vectors to be estimated, and u_2 is the disturbance vector.

The following assumptions complete the specification of the model. Denote by u the vector of stacked disturbances whose transpose is the new row vector $[u_1' \; u_2']$, and let the conditional density of u given X_1, X_2 be $f(u \mid X_1, X_2)$. Then it is assumed that:

(i) $E(u \mid X_1, X_2) = 0; E(uu' \mid X_1, X_2) = \Sigma.$

That is, conditional on X_1, X_2, the disturbance term u has a multivariate distribution with mean zero and positive definite covariance matrix Σ, where

$$\Sigma = \begin{bmatrix} \sigma_1^2 I & \sigma_{12} I \\ \sigma_{12} I & \sigma_2^2 I \end{bmatrix}$$

This specification allows the ith disturbances in the two equations to be correlated, so that unmeasured factors may influence both selection and behavior for individual i; but any other pairs of disturbances are assumed uncorrelated.

(ii) Rank $(X_1) = k_1$; rank $(X_2) = k_2$.

(iii) At least one exogenous variable in X_1 whose coefficient is nonzero is excluded from X_2.

Assumptions (ii) and (iii) guarantee that the coefficients in Equations (A-1) and (A-2) are identifiable (by virtue of meeting the usual rank condition) and that the OLS estimates set out below exist in probability.

(iv) $X_1'X_1/n$ and $X_2'X_2/n$ converge in probability to constant positive definite matrices, and $X_2'X_1/n$ converges in probability to a constant matrix.

This assumption is needed to ensure that OLS estimates of equations (A-1) and (A-2) will converge in probability.[5]

(v) The elements of u_1 and u_2 have absolute fourth moments uniformly bounded above.

This assumption guarantees, for example, that $u_1'u_1/n$ and $u_1'u_2/n$ converge in probability.

The preceding assumptions ensure that OLS estimates of Equation (A-1) are consistent. In the outcome equation (A-2), however, the OLS estimate is:

$$\begin{bmatrix} \hat{\gamma} \\ \hat{\beta}_2 \end{bmatrix} = \begin{bmatrix} y_1'y_1 & y_1'X_2 \\ X_2'y_1 & X_2'X_2 \end{bmatrix}^{-1} \begin{bmatrix} y_1'y_2 \\ X_2'y_2 \end{bmatrix} \tag{A-3}$$

Let $S^2 = y_1'(I - X_2(X_2'X_2)^{-1} X_2')y_1$, which is bounded away from zero in probability, and set

$$b = (X_2'X_2)^{-1}X_2'y_1$$

These quantities are the residual sum of squares and the regression coefficients in the regression of y_1 on X_2. Then from standard results on the inverse of a partitioned matrix:

$$\begin{bmatrix} y_1'y_1 & y_1'X_2 \\ X_2'y_1 & X_2'X_2 \end{bmatrix}^{-1} = \frac{1}{S^2} \begin{bmatrix} 1 & -b' \\ -b & S^2(X_2'X_2)^{-1} + bb' \end{bmatrix}$$

[5]Assumption (iv) may be used to show that Assumption (ii) holds in probability, which would suffice for present purposes. In that sense, Assumption (ii) is dispensable. Keeping this assumption avoids some purely technical complexities, however, and so it has been retained.

Now using

$$y_1'y_1 - b'X_2'y_1 = S^2$$

and

$$y_1'X_2 - b'X_2'X_2 = 0$$

and substituting for y_1 in Equation (A-3), we obtain

$$\begin{bmatrix} \hat{\gamma} \\ \hat{\beta}_2 \end{bmatrix} = \begin{bmatrix} \gamma + (y_1'u_2 - b'X_2'u_2)/S^2 \\ \beta_2 + (X_2'X_2)^{-1}X_2'u_1 + (bb'X_2'u_2 - by_1'u_2)/S^2 \end{bmatrix}$$

Since plim $X_2'u_2/n = 0$ and since plim S^2/n may be shown to be a positive constant, say s^2, then

$$\text{plim} \begin{bmatrix} \hat{\gamma} \\ \hat{\beta}_2 \end{bmatrix} = \begin{bmatrix} \gamma + (\text{plim } y_1'u_2/n)/s^2 \\ \beta_2 - (\text{plim } by_1'u_2/n)/s^2 \end{bmatrix} \tag{A-4}$$

But plim $y_1'u_2/n = \sigma_{12}$ and plim $b = \text{plim}(X_2'X_2/n)^{-1}(X_2'X_1/n)\beta_1 = B$, say, a constant vector. Hence

$$\text{plim} \begin{bmatrix} \hat{\gamma} \\ \hat{\beta}_2 \end{bmatrix} = \begin{bmatrix} \gamma + \sigma_{12}/s^2 \\ \beta_2 - B\sigma_{12}/s^2 \end{bmatrix} \tag{A-5}$$

Thus the estimates of both γ and β_2 are inconsistent by a quantity proportional to σ_{12}/s^2.

When y_1 is dichotomous (1 = assignment to treatment group, for example, and 0 = assignment to control group), Equation (A-4) again applies. In the resulting probability limit, however, it is now no longer the case that σ_{12} is fixed. Since y_1 is dichotomous, so are its disturbances, and it is well known that the constant variances and covariances previously assumed for the disturbances are inappropriate for this case. That is, Σ must be modified.

To get some insight into the behavior of the inconsistency in this case, suppose that y_1 is generated from a probit specification. That is, for some unobserved y_1^*

$$y_1^* = X_1\beta_1 + u_1^* \tag{A-6}$$

where all terms are defined by obvious analogy to (A-1), assumptions (i)-(iv) hold, and the elements of u_1 are assumed to be distributed independently and normally with mean zero and unit variance. Then it is assumed that individual elements of the vector y_1 are 0 or 1 according to whether y_1^* is negative or nonnegative, respectively. With additional technical assumptions, these conditions guarantee that maximum-likelihood estimates of (A-6) exist and are consistent (Haberman, 1974, pp. 314-21).

It will be assumed that u_1^* and u_2 are jointly normally distributed with covariance matrix $\sigma_{12}^* I$. It follows from standard results on the bivariate normal distribution (Lord and Novick, 1968, pp. 337–44) that the covariance for the ith observation is $\varphi_i \sigma_{12}^*$, where φ_i is the ordinate of the unit normal density at $X_{1i}\beta_1$ and X_{1i} is the ith row of X_1. This result modifies the inconsistency in (A-5) for $\hat{\gamma}$ to $\sigma_{12}^* \varphi / s^2$, where $\varphi = \Sigma \varphi_1 / n$. The inconsistency for $\hat{\beta}_2$ is adjusted similarly.

This new result shows that in the case of dichotomous selection, the inconsistency behaves essentially as it did in the continuous case. The numerator is again the mean covariance between the disturbances; the difference now is that this quantity varies by observation. The inconsistency vanishes as the numerator goes to zero, just as before. If there are no excluded variables common to both disturbance terms u_1^* and u_2, for example, then σ_{12}^* will be zero and no inconsistency will occur. Alternatively, if selection is made by some mechanical rule and can be predicted nearly perfectly, $X_{1i}\beta_1$ will always be a forecast several standard deviations above or below zero on the normal abscissa (meaning a probability of joining the treatment group that is nearly 1 or 0), and hence the height of the normal density for each observation will be nearly zero, making φ nearly zero. Again this selection method essentially eliminates inconsistency, just as in the continuous case. The sole difference is that in the continuous case σ_{12} goes to zero when prediction improves in the selection equation, whereas in the dichotomous case σ_{12}^* is assumed fixed and it is φ that approaches zero. The effect is the same.

This analysis of the dichotomous case extends in an obvious way to the situation in which only the outcome variable is dichotomous (1 = success; 0 = failure) and to the case in which both selection and behavior variables are dichotomous. The results in Equation (A-4) apply, so that inconsistencies in terms of the probit parameters can be obtained simply by evaluating the two expressions, s^2 and plim $y_1' u_2 / n$, in the manner of the immediately preceding paragraphs.

3

Estimating Treatment Effects in Quasi-Experiments: The Case of Nonrandomized Assignment

INTRODUCTION

The previous chapter demonstrated that simple statistical techniques, such as contingency tables or regression analysis, do not yield appropriate estimates of treatment effects in quasi-experiments. Comparing "matched" members of treatment and control groups also fails. An accurate assessment of a quasi-experiment depends on explicit modeling of both the behavioral outcome of the experiment *and* the assignment to treatment groups. When this has been done, statistical methods from the theory of simultaneous equations may then be used to give dependable estimates of the treatment's impact. This chapter sets out the mechanics of these estimators in the simplest cases. More complex cases are deferred to Chapter 6; proofs are relegated to the appendix.

Consider first the simplest (two-equation triangular) case, in which both the treatment and the experimental outcome are continuous variables. The first equation explains the choice of a treatment (the assignment equation); the second accounts for the result of the treatment (the outcome equation). Treatments influence outcomes, but not vice versa. In the simplified Coleman Report model considered in the previous chapter, for example, the fraction of middle-class students in a child's school was the treatment and achievement test scores were the outcome under study:

$$\text{School Percentage Middle Class} = a_1 + b_1\,(\text{Middle Class}) + b_2\,(\text{Urban}) + u_1 \tag{1}$$

$$\text{Achievement} = a_2 + b_3\,(\text{School Percentage Middle Class}) + b_4\,(\text{Middle Class}) + u_2 \tag{2}$$

In studies like this one, it is often reasonable to assume that, taken together, the unobserved factors influencing both school choice and test results generate essentially random deviations from what would otherwise have occurred—and, moreover, that these deviations are not expected to be systematically larger for one child than for another. More precisely, it is usually assumed in such cases that the disturbance terms in the two structural equations have mean zero, that they are uncorrelated with the exogenous variables (variables other than assignment and outcome), and that their variances and covariance with each other are constant across observations (homoskedasticity). The matrix of observations on the exogenous variables is also assumed to have full rank. (Perfect collinearity among the exogenous variables is excluded.)

In addition to these assumptions, one more condition must be met before statistical theory can disentangle a quasi-experimental effect from the assignment procedure. There must be at least one *exclusion restriction* in the equation that models behavior. That is, at least one variable must be found that influences assignment but does not influence the experimental outcome. The variable Urban serves this function in the Coleman example. To say that such a variable has been found is to claim that if the excluded variable were added to the outcome equation, its true coefficient would be zero.

It is sometimes thought that if a variable is known to have no true *causal* effect on outcomes (but some impact on assignment), then it qualifies as an excluded factor. If models are properly specified causally, this logic is correct: having no causal effect is equivalent to having a zero coefficient. In practice, however, social science models are always improperly specified in a causal sense, sometimes deliberately so. The genuine causes of human behavior are too varied to measure accurately, and proxy variables are frequently used to take the place of unmeasured causes. For example, researchers may control for a respondent's city size or region of the country in predicting his purchases, his vote, or his school achievement, not because place of residence directly causes behavior but because it summarizes the attitudes, opportunities, and social networks in which the respondent is likely to find himself. Place of residence is a proxy—a deliberate causal misspecification to avoid the staggering costs of gathering more accurate data. The resulting equation may be well specified statistically (unmeasured factors in the disturbance term uncorrelated with exogenous variables), but the coefficients no longer represent true causal impacts. In particular, some variables with no genuine effects will have large coefficients.

Excluded variables in a simultaneous equation must have a coefficient of zero. In practice, this means that the variables must have neither a causal impact of their own nor a proxy effect for other excluded variables. It is not enough that they do not directly influence the behavior of interest; they must not be correlated with variables that do.

In some instances, variables of this kind are easy to find. If a school district draws its boundaries along the middle of ordinary residential streets, it seems reasonable to suppose that there will be no systematic differences in the families on opposite sides. Presumably, therefore, the child's address will influence assignment but not behavior and will not be correlated with other causes of behavior. Hence street address is an excellent excluded variable in the outcome equation. In other instances, finding such a variable is much less simple and factors thought to have small effects (rather than none) are chosen instead. The more uncertain the impact of the excluded variables, the more dubious the ensuing results. In some studies, the search for suitable excluded variables may be exceptionally difficult. Nevertheless, the need for exclusions cannot be circumvented in quasi-experiments.[1]

---

[1]One can proceed logically from evidence to conclusion only with the help of randomization or prior substantive knowledge. One or the other must be supplied. In the absence of randomization, the requirement that the outcome equation be identified cannot be escaped. Assuming that nothing is known a priori about variances or covariances of the disturbance terms—the usual situation—then something must be known about the coefficients. In practice, this requirement is nearly always met by assuming that one or more coefficients are zero. Hence exclusion restrictions are the only practical alternative to randomization.

THE ASSIGNMENT EQUATION: OLS, GLS, AND PROBIT

We now consider the estimation of a quasi-experiment with linear assignment and outcome equations. First, assume that the assignment equation is

$$y_1 = a_1 + b_{11}x_1 + b_{21}x_2 + \cdots + b_{k1}x_k + u_1 \qquad (3)$$

where y_1 is the treatment, the x_j's are exogenous factors influencing the assignment to treatments, a_1 is an intercept, the b_{j1}'s are coefficients, and u_1 is the disturbance term. If y_1 is a continuous variable, this expression has the form of an ordinary regression specification. As is well known, under the assumptions listed below equation (2), OLS is an unbiased estimator of the coefficients, and it has minimum variance in the class of unbiased linear estimators. Under mild additional conditions it is consistent as well. In general, the first equation in a triangular simultaneous-equation system may be estimated by OLS with no special econometric difficulties. Ordinary regression assumptions suffice.

In many quasi-experiments, however, y_1 is not continuous. When subjects are assigned to either a treatment or a control group, for example, y_1 can take on just two values, say 0 (control) or 1 (treatment). In this situation, one popular approach is to model the process generating the assignments as a regression equation. The dichotomous treatments are made the dependent variable, and they are regressed linearly on the factors thought to influence them.

The true forecast values from a regression of this kind, which are assumed to fall between 0 and 1, are interpreted as probabilities. A forecast of 0.7, for example, means that an observation has a 70 percent chance of equaling 1 (assignment to the treatment group) and a 30 percent chance of 0 (assignment to the control). Since these probabilities are linear in the independent variables, this version of regression is called the *linear probability model*.

Because the dependent variable is dichotomous, the disturbances in the linear probability model do not have the constant variance that is required by the regression procedure; they are *heteroskedastic* (Goldberger, 1964, pp. 248–50). If no other violations of the assumptions are present, the OLS coefficient estimates are unbiased but their standard errors are incorrect. The appropriate standard errors can be obtained only by repeating the computations with suitable weights for the observations. These second-round coefficients and standard errors are called *generalized least squares estimates* (GLS) or "Goldbergerized estimates." The computations for the linear probability model thus require two rounds of regression:

A. Apply ordinary regression with the dichotomous treatment as the dependent variable.
B. For each observation, let the forecast value be p and adjust forecasts above 0.99 to 0.99 and those below 0.01 to 0.01. Set $q = 1 - p$ and $s = \sqrt{pq}$.
C. Divide each variable in the regression by s. The intercept term becomes $1/s$. Apply ordinary regression to this new set of variables, including the variable $1/s$ but suppressing the conventional intercept. The resulting coefficients and standard errors are the linear probability model estimates.

An alternative to the linear probability technique is the probit model (Finney, 1947). If subjects are assigned to either a treatment or a control group, then the assignment equation governing that choice may be specified as a probit function of the exogenous variables. This function was introduced formally in the appendix to the previous chapter and will be explained briefly here in more intuitive terms.

Like the linear probability model, probit analysis makes the dependent variable the probability of being assigned to the treatment group. However, the probit specification assumes nonlinearity: a unit change in variables on the right-hand side produces a larger change in this probability if subjects start from a 50 percent chance than if they start from a 90 percent chance or a 10 percent chance. Near 0 and 100 percent, change becomes very difficult, so that no one is ever predicted to have less than 0 percent or more than 100 percent chance of assignment to the treatment. This characteristic of probit analysis contrasts with the dichotomous regression or linear probability case, in which forecasts outside the 0 to 1 interval are not uncommon, and change in probability is assumed to be generated in equal amounts no matter what the initial probability of assignment. For this reason, the probit model often fits the data somewhat better than the linear probability model, especially if the observations are spread fairly evenly over the entire range of probabilities.

Although the nonlinear relationship between independent and dependent variables in probit analysis often improves the specification, it also makes the interpretation of the coefficients more subtle. Probit analysis assumes that the effects of the independent variables become linear and additive when the original assignment probabilities are nonlinearly transformed to a different scale. Formally, in the probit model, the probability that an individual will have the value 1 on the dependent variable is found by computing the linear additive forecast value from the probit coefficients and then finding the area to the left of that point on a standard normal curve. In practice, a good rule of thumb is that for individuals whose estimated probability of assignment to the treatment group is, say, 25 to 75 percent, dividing a probit coefficient by 3 will give an approximation to its effect on the probability. Thus for these individuals, a probit coefficient of 0.30 would correspond to a linear probability coefficient of roughly 0.10. In both cases, a one-unit increase in the variable results in an increase of about 0.10 (ten percentage points) in the probability of assignment to the treatment group. (In fact, the marginal effect of the probit coefficient would range from 0.09 to 0.12 over the 25 to 75 percent interval.) For individuals with estimated probabilities farther from 50 percent, of course, a probit coefficient will have less effect than 0.10.

When the treatment variable, y_1, is generated by a probit specification, ordinary probit analysis using the estimation procedure known as maximum likelihood gives consistent estimates of the probit coefficients in the assignment equation. The requisite assumptions closely resemble those for consistent regression analysis. (See the discussion of Equation (A-6) in the appendix of Chapter 2.) The software for this procedure is widely available.

THE OUTCOME EQUATION: 2SLS AND G2SLS

When at least one exclusion restriction can be established with confidence and when treatment and outcome are each continuous variables, then *two-stage least squares* (2SLS) is the standard estimation procedure for the outcome equation. Software to carry out the computations is included in many statistical programming packages, but the estimates can also be computed by using an OLS routine.

Assume that in the quasi-experiment to be analyzed, the outcome equation is of the form

$$y_2 = a_2 + cy_1 + b_{12}x_1 + \cdots + b_{K2}x_K + u_2 \tag{4}$$

where y_2 is the outcome, y_1 is the treatment, the x_j's are exogenous control factors, not necessarily the same as those in equation (3), and a_2, c, and the b_{j2}'s are coefficients to be estimated. Assume also that the probability of being in the treatment group is solely a function of certain exogenous variables. (For the time being, it is not important to specify the form of the assignment equation; for example, it need not be equation (3). This point is discussed in more detail below.)

In equation (4), the elements of the disturbance vector u_2 are assumed to have mean zero, constant variance, and zero covariances, each conditional on the x_j's. It is also postulated that there is at least one exogenous variable (uncorrelated with u_2) from the assignment equation that does not appear among the x_j's in equation (4). The complete collection of exogenous variables from the assignment and outcome equations constitute the exogenous variables in the model. The corresponding matrix is assumed to have full rank to ensure that the 2SLS estimates exist.

Two-stage least squares may be carried out as follows:

A. Regress the treatment variable on all the exogenous factors in the model. This step constitutes the first-stage regression.
B. Replace the original treatment variable y_1 in equation (4) with the forecast ("purged") values from the regression (step A). Apply ordinary regression to this new equation (the second-stage regression). The resulting coefficients are the 2SLS estimates. With some plausible additional conditions of a technical nature (see the appendix), these estimates are consistent and asymptotically normally distributed.
C. Let the variance of the residuals from the second-stage regression be ω^2. Create a new set of residuals for the same equation using the same coefficients, but replace the purged treatment variable with the original unpurged variable. Compute the variance of the new residuals and denote it by σ^2. Multiply each standard error of a coefficient as computed in the second stage by $\sqrt{\sigma^2/\omega^2}$. The result is the appropriate 2SLS standard errors of the coefficients.

The 2SLS procedure is designed for continuous outcomes. In many policy experiments and elsewhere in the social sciences, however, outcomes are dichotomous—success or failure, bought or did not buy, voted Democratic or voted Republican. In a quasi-experiment with a dichotomous outcome variable, the disturbances in the outcome equation are heteroskedastic. This means that if a simultaneous-equation estimator is needed, 2SLS is inappropriate since it assumes that the variance of the disturbances is constant. If the other assumptions for 2SLS hold, then with additional technical constraints, the effect of ignoring the dichotomous nature of the outcomes closely parallels that in the ordinary regression case: the coefficient estimates are consistent, but their standard errors are erroneous.

Fortunately, 2SLS can be extended to the dichotomous endogenous variable case in much the same way that regression is generalized in the linear probability model. One simply codes the outcomes as 0 or 1 and assumes that the probability of observing a 1 is linearly related to the right-hand-side variables. The new estimator of this equation can be referred to as *generalized two-stage least squares* or *G2SLS*. The computational method is as follows:

A. Apply 2SLS. (The resulting coefficient estimates will be consistent, but the standard errors will be wrong.) Next compute the forecast value of the dichotomous left-hand variable in the outcome equation, using the second-stage coefficients and the *original* (unpurged) right-hand variables. For each observation, denote the predicted value by p, denote $(1 - p)$ by q, and let $s = \sqrt{pq}$. If p exceeds 0.99 or falls below 0.01, it should be reset to those bounds so that s exists and is bounded.

B. Create new variables by dividing every variable, including the dichotomous one and the exogenous variables excluded from the outcome equation, by s. If the original outcome equation contained a constant term (the intercept), the corresponding new variable will be $1/s$. Apply 2SLS to this new set of variables, suppressing the conventional intercept in both the first and second-stage regressions.[2] The resulting coefficient estimates are the G2SLS estimates.

[2]It bears emphasizing that in purging a dichotomous variable, forecasts outside the range 0.01 to 0.99 are *not* reset to the endpoints. Doing so alters the standard errors of the coefficients. Only when these forecasts are used in computing a standard deviation s are they restricted to lie in this interval.

C. Let the residual variance from the final regression in step 2 be ω^2. Then if each coefficient standard error is multiplied by $\sqrt{1/\omega^2}$, the appropriate standard errors will result.

When the outcome variable is dichotomous and generated by a linear probability specification, then under standard assumptions, this procedure is more efficient than 2SLS—that is, it yields coefficients whose standard errors are asymptotically smaller. In fact, within the wide class of techniques known as *single-equation instrumental variable estimators*, G2SLS is fully efficient asymptotically. As the sample size tends to infinity, this estimator gives the coefficients with the least sampling error in its class. Moreover, under mild additional conditions these coefficients are asymptotically normally distributed. (This estimator is developed formally in the appendix in the context of a more general model.)

Under its assumptions, G2SLS is suitable whenever the outcome variable is dichotomous, regardless of whether the treatment variable is continuous or dichotomous. The same calculations apply in either case. Extensions of this procedure to cope with other sorts of heteroskedasticity are discussed in Chapter 6.

Finally, note that when both the assignment and the outcome equations are specified to be linear probability models, the procedure for estimating the assignment equation is similar to the purging steps in G2SLS. Both require two rounds of regression, with the second round being a heteroskedasticity correction. The principal difference is that the standard deviation s in the linear probability model for the assignment equation is computed from the residuals in *that same* equation, whereas the corresponding s throughout the G2SLS computations is the standard deviation of the residuals in the *outcome* equation. Consequently the final estimates from the assignment equation cannot be used to compute forecasts for the purging step in the G2SLS estimates of the outcome equation. (Such an estimator is consistent, but its standard errors are not those produced by the G2SLS calculations, and it is less efficient than G2SLS as well.)

NONLINEAR OUTCOME EQUATION

A variety of nonlinearities can be handled within the preceding framework. First, certain apparently nonlinear equations can be transformed to a linear specification—for example, by taking logs of both sides. The usual estimation procedures, such as 2SLS, may then be applied. This trivial case raises no new questions, and it need not be discussed further.

Second, certain structural equations are nonlinear due to the presence of one or more nonlinear terms on the right-hand side. For instance, one of the exogenous factors influencing outcomes in a policy experiment might be x_1^2 or an interaction term like x_1x_2. So long as these factors enter the equation additively—and are functions only of the *exogenous* variables—once again no new principles arise. Each such nonlinear term is treated as just another exogenous variable, and estimation proceeds as before. If x_1, x_2, and x_1x_2 are the exogenous variables appearing in the assignment and outcome equations, for example, then to apply 2SLS one would purge the treatment variable with all three. In every other respect, 2SLS would be executed exactly as in the purely linear case. Consistent estimates and appropriate standard errors will result.

When nonlinearities involving *endogenous* variables appear on the right-hand side, more care is required, but again simple extensions of 2SLS will produce attractive estimates. It is required only that the equations be linear in parameters—that is, they must be expressable in a form like (4), with y's and x's on the right-hand side replaced by *known* functions of these same variables. These functions cannot involve any unknown coefficients: $x_1^2y_2$ is a suitable nonlinear variable; y_1^γ is not. Nonlinear equations of this kind are said to be *linear in the parameters*.

One consistent estimate of nonlinear equations which are linear in parameters is Kelejian's (1971) nonlinear two-stage least-squares (NL2SLS) estimator. Assuming that the exogenous variables are *independent* of the disturbances (not merely uncorrelated), the researcher first selects low-order polynomial terms in the exogenous variables that are believed to predict the nonlinear functions on the right-hand side of the outcome equation. These polynomial variables are all counted as additional excluded exogenous variables. If the outcome variable is continuous and meets the usual assumptions, ordinary 2SLS is then applied; the purging is carried out by using this augmented set of exogenous variables. This procedure is consistent and gives the appropriate standard errors, although it can be quite inefficient if the nonlinear functions are not well predicted by the full set of exogenous variables.

Suppose, for example, that the outcome equation has the quadratic form

$$y_2 = a_2 + cy_1^2 + u_2 \quad (5)$$

and assume that x_1 and x_2 are exogenous variables appearing in the assignment equation. Then to apply NL2SLS, one might regress y_1^2 on x_1, x_2, and perhaps x_1^2, x_2^2, and x_1x_2 as well. Forecast values of y_1^2 would then be inserted into the outcome equation (5) in place of y_1 itself, and OLS would be applied. With the usual adjustment of the standard errors afterward, this second step would provide consistent estimates of the parameters in equation (5), namely a_2 and c, along with their standard errors.

Notice that one may not estimate this equation by purging y_1, squaring the resulting forecasts, and inserting them into equation (5) to run OLS. This method is inconsistent (Kelejian, 1971). In general, nonlinear simultaneous equations cannot be estimated by purging the endogenous variables separately and then inserting the purged values into nonlinear functions. Instead, the nonlinear functions themselves must be purged.

A more efficient solution than NL2SLS can be obtained in certain cases. Consider, for example, the two-equation quasi-experiment specification in which y_2 is a simple linear function of y_1 and exogenous variables as in equation (4) but y_1 is known to be generated by a probit setup in the assignment equation. This latter equation can be estimated by ordinary probit analysis. Forecasts of y_1 may then be obtained by computing $\Phi(y_1^*)$, where y_1^* is the forecast based on the probit coefficients (the *normit*) and $\Phi(\cdot)$ is the cumulative distribution function of a unit normal curve, that is, the area to the left of y_1^* under a normal curve with mean zero and variance 1.[3]

Next a modified version of 2SLS is used to estimate the assignment equation. First, the purging variables to be used in estimating the

[3]This function is available in much social science software, and a reasonably accurate, simple approximation to it is given by $(G + 1)/2$, where $G = [1 - \exp(-2y_i^{*2}/\pi)]^{1/2}$ and $\exp(z) = e^z$ (Kendall and Stuart, 1969, vol. 1, p. 367).

outcome equation are $\Phi(y_1^*)$ plus those exogenous variables explicitly included in the outcome equation. (In contrast to ordinary 2SLS, exogenous variables that appear in the selection equation but not in the outcome equation are *not* used.) Then 2SLS is applied, using the purging variables as regressors in the first stage. The rest of the computations are carried out just as before, giving estimates that are asymptotically more efficient than those of nonlinear 2SLS and fully efficient in the class of single-equation instrumental variable estimators. If the outcome equation is heteroskedastic, G2SLS using the modified list of first-stage regressors is a suitable estimator just as in the homoskedastic case. (For proofs, see the appendix.)

Although the foregoing estimator is efficient, it does require knowledge of the functional form of the assignment equation. In many instances, however, the assignment equation is not of much interest. In that case, Kelejian's NL2SLS technique can be applied—for example, by purging the treatment variable by ordinary linear regression on all the exogenous variables in the system. Then in the outcome equation 2SLS or G2SLS will produce correct (though perhaps inefficient) coefficient estimates and standard errors regardless. For instance, if the true selection equation takes the probit form but the dichotomous treatment variable is purged with ordinary regression, the resulting coefficients and standard errors in the outcome equation will remain correct. More efficient estimates could be obtained in this case by using the G2SLS estimator outlined above, but this procedure requires knowledge of the true functional form of the selection equation, which may not be known with any confidence.

Other nonlinearities in a structural equation can also be handled by methods like those described in this chapter, so long as the parameters enter linearly. These extensions are taken up in Chapter 6.

PROBIT ANALYSIS IN THE OUTCOME EQUATION

When the policy outcome is dichotomous, a researcher may wish to model it as a probit equation. If the treatment variable is continuous and the assignment equation is linear with homoskedastic normally distributed disturbances, methods of estimation resembling 2SLS have been proposed by Maddala and Lee (1976), Heckman (1978), and Amemiya (1978). A good overview is Maddala (1983). The simplest method parallels 2SLS closely, with the addition of a coefficient-rescaling step at the end:

A. Purge the treatment variable on all the exogenous variables in the model.
B. In the outcome equation, replace the original values of the treatment variable with the purged values. Apply probit analysis.
C. Divide each resulting probit coefficient by s, the square root of s^2, where $s^2 = 1 + \text{var}(\hat{y}) - \text{var}(\tilde{y})$. Here $\hat{y}$ is the probit forecast from step (B), $\tilde{y}$ is the probit forecast using the same coefficients but with the original values used instead of the purged values, and the variances are computed over the sample. The resulting probit coefficients are consistent, although the standard errors are wrong and cannot be corrected simply.

This method will also work if the assignment process is a probit equation, so long as the treatment is the underlying unobserved continuous variable rather than the dichotomy or polychotomy that is actually observed. Thus, for example, survey researchers typically ask respondents to match their opinions to one of a few discrete categories. However, the actual opinion is presumably not discrete, but a point on an underlying continuous scale. Similarly, in a quasi-experiment the actual treatment may be continuous but measured only on an ordinal scale. In such cases, polychotomous probit analysis may be used to estimate the assignment equation. The predicted values from this equation constitute purged values of the underlying continuous variable, and they may be inserted into the outcome equation to produce estimates that would be consistent if rescaled appropriately. Unfortunately, neither the scale nor the standard errors can be easily corrected. The relative magnitudes of coefficients may legitimately be compared, however.

The case just discussed is somewhat atypical. When the assignment equation has a probit specification, the treatment is usually not the underlying continuous variable, but rather the observed dichotomy. This case occurs when subjects are assigned to treatment or control groups, for example. Then none of the relatively simple methods discussed in this chapter is appropriate. Ignoring the dichotomy and applying any of the techniques suited to continuous variables will produce inconsistent estimates. The reason is that under all estimation methods like 2SLS, purging the treatment variable in effect causes the disturbances from the assignment equation to be added to the probit disturbances. But since the treatment variable is dichotomous, its disturbances from the assignment equation cannot be normal. Hence adding them to the probit disturbances produces nonnormal disturbances. Probit analysis is inconsistent in general when its disturbances are not normally distributed.

These difficulties make probit estimation in the outcome equation rather troublesome for applied work. However, computational procedures for obtaining coefficient estimates and standard errors in a wide variety of cases not considered here may be found in the references cited above.

AN EXAMPLE OF G2SLS

The discussion so far has made the standard econometric assumptions that the data set is fixed, the true functional form of the model is known, and the variables have been selected in advance. In practice, none of these assumptions hold. A single data set must be selected from among alternatives, bad or irrelevant data points must be pruned away, and good specifications must be found by trial and error, with no one of them being unequivocally best. In such circumstances, drawing dependable inferences and presenting them honestly demands skills whose rules cannot be inferred from econometric theory. In learning from quasi-experiments, substantive understanding and careful data analysis proceed side by side with statistical knowledge. To illustrate the nature of the enterprise, an example is now given from the study of pretrial release.

Within a day or two after a defendant is arrested, he or she is brought before a judge to determine eligibility for release. In most jurisdictions, the judge has two principal options available: OR (release on own recognizance) or bail (release only if the defendant can post a bond for his return). Both options have many variations, of course, including release to hospitals for drug or alcohol evaluation and treatment, release supervised by another person or agency, cash bail (which requires only a small fraction, usually 10 percent, of the original bail amount), and a variety of other possibilities. Nevertheless, the fundamental judicial choice is between OR and bail.

Defendants who neither receive OR nor post bail are held in jails. Since trials are often delayed, detention in many jurisdictions results in terms of confinement extending over weeks and months. Jails are local institutions, and life in them is often hideous even by the modest standards of state and federal prisons. Thus defendants in pretrial detention may be subjected to more rigorous punishment than they would receive if convicted—even though they are constitutionally innocent until tried.

The conditions of American jails have sparked periodic crusades for reform (Beeley, 1927; Freed and Wald, 1964; U.S. Dept. of Justice, 1965; Executive Board of the National Conference on Bail and Criminal Justice, 1965). Some institutions have better facilities as a result, and many jurisdictions provide more pretrial information for judges and wider use of OR. Some researchers, however, have argued that detention itself is constitutionally problematic. Beginning with Morse and Beattie (1932), they have claimed that pretrial incarceration increases the accused's chances of being judged guilty and sent to jail. Foote (1954, p. 1052) found, for example, that in Philadelphia during October and November 1953, some 72 percent of jailed defendants, but only 52 percent of those released, were ultimately convicted (see also Friedland, 1965). Confinement, it is said, restricts the defendant's opportunities to prepare a defense. He cannot interview witnesses, raise funds, or stay out of trouble before trial as a demonstration of his innocence. Moreover, at trial the detained defendant enters the courtroom from jail in the company of a bailiff, thereby influencing judge or jury to think him guilty. The result is said to be a higher rate of conviction and jail sentences among defendants held before trial.

The alleged invidious effect of detention on trial outcomes is particularly important from a legal point of view. For if detention per se prejudices a defense, the case can be made that detention is unconstitutional, at least under current conditions. Pretrial reform—either more releases or better conditions for those detained—then may be available through the courts.

Precisely this argument was made in the early 1970s by the Legal Aid Society of New York City (1972) in *Bellamy v. Judges*. The society lost its case, however, at least partly for the right reason. Any comparison of rates of guilt or jail sentences between jailed and released defendants constitutes a quasi-experiment. No jurisdiction assigns the accused to jail or freedom randomly; rather judges attempt to hold those who represent a risk to society, either because they may flee the jurisdiction to avoid trial or more often, because they represent a danger to society on release. Most state bail laws explicitly instruct judges to evaluate the strength of the evidence, the probability that the defendant will flee, and, increasingly, "dangerousness," as indicated by the seriousness of the charge, prior record, and so on. In any case, political realities force judges to take all these factors into account. It follows, then, that detained and released defendants constitute very different groups of people.

A showing that detained individuals are more likely to be convicted and sentenced to prison thus proves very little. While it is true that this outcome is to be expected if detention is discriminatory, it is also the expected result if detention has no effect. So long as judges perform their tasks in general compliance with the law, detained defendants are more likely than those on release to be charged with serious crimes and they are more likely to be guilty. Hence they are convicted and sentenced to prison terms more often, regardless of whether detention itself has an effect.

Researchers in the field sometimes recognize this argument, but they appeal to the size of the disparities in conviction and sentencing rates as evidence for their position. Thus Foote (1954, p. 1054) argues that "the contrast in comparative dispositions was so striking that it is reasonable to conclude that jail status had a good deal to do with it." But of course this argument is fallacious. The fact that an effect is large provides no evidence whatever that a particular factor caused some of it. The bias in comparing raw differences in conviction and sentencing rates is of the sort discussed in Chapter 2, where it was shown that substantial differences between quasi-experimental treatment and control groups can be created with no treatment effect at all. In a pretrial release system, good judicial forecasting will produce large differences in conviction and incarceration rates, even in the absence of any discrimination. This simple argument tells against the entire body of research on which the case for pretrial discrimination depends.

In the language of quasi-experimentation, the "treatment" here is pretrial detention and the "outcome" is the trial judgment. Assignment to treatments is nonrandom; defendants are chosen for release in part because they are expected to do well in court. Many of the same factors that save an accused criminal from detention will also save him from conviction and incarceration. Because researchers cannot measure all of these elements (strength of the evidence, defendant's court demeanor, caliber of his attorney), they relegate them to the disturbance term in both the selection and outcome equations. The disturbances then are correlated, and ordinary cross-tabulation and regression fail. Assessing the true impact of pretrial detention on the probability of conviction and a prison sentence requires a statistical model of the kind discussed earlier in this chapter.

A data set to which the model may be applied was collected by the District of Columbia Bail Agency in 1976. (A complete discussion of the data is given in Welsh and Viets, 1976.) It contains a wide variety of information on each serious misdemeanor or felony defendant in Washington, D.C., in the calendar year 1975. It records whether defendants received OR or bail, whether they were ultimately adjudged guilty or innocent, and, if guilty, whether they were sentenced to prison. The original data did not contain information on the release status of defendants assigned bail; however, for the purposes of this study, the relevant information was drawn from court files in the District of Columbia and added to the data set.

The Bail Agency sample excludes traffic offenses, violations, and crimes charged by the Corporation Counsel, along with juvenile offenses not charged as adult crimes. Individuals not interviewed by the Bail Agency (about 5 percent of all defendants) were also dropped (Welsh and Viets, 1976, p. 20). This left 20,109 cases—a case being defined as a set of related charges brought against a single defendant. (Individuals working together generate separate cases; so does a single individual who is arrested more than once during the year.) The analysis reported here restricted the sample still further to obtain a more meaningful and homogeneous group of defendants amenable to statistical analysis. Defendants who departed from the system before their pretrial release hearing were excluded; this group includes those who pled guilty or had their cases dismissed at arraignment, as well as those who were diverted to alcohol or drug treatment programs. Furthermore, only adult cases in superior court and without political implications were included. All juveniles, all district court defendants, and all "special processing" cases (diplomats, sons of members of Congress, and so on) were excluded. Also excluded by necessity were those defendants with missing data for one or more of the relevant variables—for example, those who skipped the jurisdiction, those whose cases were still in process when the data were collected, or those whose court records were lost or unreadable. The remaining sample for the analysis reported here constituted approximately 11,000 observations. (The number varies slightly depending on which variables are used.)

Pretrial detention is said to affect both conviction and incarceration rates. In practice, conviction is much easier to study than incarceration. In Washington as elsewhere, defendants are often sentenced to "time served (in pretrial detention)." In other cases, a sentence of some months is given, but the records do not specify whether the time already served is to be applied toward it. Of course, a defendant receiving an explicit or implicit sentence of time served is released immediately. Thus individuals detained before trial and likely to be judged guilty have no incentive to resist plea bargains in which they receive a sentence of time served, even if no such sentence would have been imposed in a trial. Prosecutors also favor such agreements, since they count as successful prosecutions and imprisonments. But no more prison time may be served than if the defendant were judged innocent. It follows, then, that legitimate sentences and those that merely paper over a pretrial detention are hopelessly confused in a typical pretrial data set such as the one at hand. Though they stem from very different causes, both result in a recorded prison sentence for the defendant. Since released defendants are never sentenced to time served, detained defendants will always appear to receive more prison sentences, even with nonrandom assignment corrected. (The effect appears in the Washington data, for example.) But this difference does not imply that detention increases the probability of incarceration (except in the tautological sense).

The inferential difficulties are considerably lessened in the case of conviction rates, however. Defendants may be tempted to plead guilty to get out of pretrial detention, but that decision is not nearly so inconsequential as agreeing to a sentence of time served. Detainees with a chance of proving their innocence have strong incentives to persist in doing so—especially if, as was true in Washington at this period, they cannot be held more than sixty days (thirty days for misdemeanors). For this reason, the empirical analysis reported here makes guilty verdicts (on the most serious charge) the dependent variable.

A proper assessment of the impact of pretrial detention requires not only a suitable statistical technique but also an appropriate set of control variables. Three kinds of controls are needed: the first is the nature of the charge (some criminal charges are easier to prove than others, and more serious charges are both prosecuted more diligently and harder to prove); the second is the resources available to the defendant (education, income); the third is the character of the defendant (prior record, race, sex). Variables fitting any of these descriptions were examined as potential control factors. First their bivariate relationship to guilt was explored. Those with nonlinear effects (such as number of prior convictions) were entered into the outcome equation with suitable transformations, either by adding a quadratic term or by replacing the linear term with a quadratic. A few dummy variables for classes of crimes proved useless (such as the felony/misdemeanor distinction); these were dropped. The remaining variables constituted the list of control factors. Other variables collected by the Bail Agency, such as age, were thought to affect the prospects of release but not to influence the likelihood of a guilty verdict. They were used as purging variables.

The other explanatory variable used was the treatment itself: pretrial release. This variable is defined by the defendant's *initial* status. That is, a defendant is "on release" if he gains release on recognizance or makes bail at his first bail hearing; he is "detained" if he is assigned a bail amount that he cannot make at this date. Some defendants initially detained later gain release, and others initially free violate the terms of their release and are remanded to detention. In fact, in long trials defendants can go to and from jail several times. Since it is impossible to follow these changes in the Washington court records, the initial release status defines the variable.

The incompleteness in the release records introduces noise into the treatment variable. Some defendants coded as detained were on release most of their period before trial, and a few initially on release were detained. Since the latter group is small, the principal effect of these errors is to contaminate the nominally detained group with a group of partly released/partly detained defendants. Although the issue is little discussed in the literature, it arises in every jurisdiction. "Pretrial release" is not a perfect dichotomy. The solution is simply to be precise about how the mixed cases are to be handled. Ignoring the few defendants who gained release only to be remanded later, the coding used here in effect counts as detained all defendants who were *ever* detained whereas the released group is everyone *never* detained. With conventional justice system data, some such definition is unavoidable; and properly understood it causes no statistical difficulties. (Precise definitions of all variables are given in Table 1.)

TABLE 1. VARIABLE DEFINITIONS FOR TABLES 2 AND 3.

	Explained (Left-Hand-Side Endogenous) Variable	
Guilt	1	If defendant was judged guilty on most serious charge
	0	If case was completed without guilty verdict on most serious charge
	missing	If case is not complete (for example, still in process or defendant skipped) or outcome is not recorded.
	Explanatory (Right-Hand-Side) Endogenous Variable	
Release	1	If defendant was released at bail hearing (either on own recognizance or by posting bond)
	0	Otherwise
	Explanatory (Included) Exogenous Variables	
	DEFENDANT'S PERSONAL CHARACTERISTICS	
Black	1	Black
	0	Otherwise
Male	1	Male
	0	Female

Variable	Code	Description
Prior Convictions		Number of prior convictions in D.C. police records, excluding traffic offenses and D.C. municipal violations (8 = 8 or more)

NATURE OF THE CHARGE

Variable	Code	Description
Violent Offense	1	If most serious charge is "violent" under Sec. 23-1331 (3) of the Bail Reform Act
	0	Otherwise (not dangerous or insufficient information)
Dangerous Offense	1	If most serious charge is "dangerous" under Sec. 23-1331 (4) of the Bail Reform Act (essentially either "violent" or drug-related)
	0	Otherwise
Crimes Against Person	1	If most serious charge is crime against person (all categories of homicide, kidnapping, sexual assault, robbery, child cruelty)
	0	Otherwise
Crimes Against Property	1	If crimes against property (all categories of arson, burglary, larceny, extortion, fraud, embezzlement, forgery, stolen property, stolen vehicles, and miscellaneous, including obstruction of mail and cruelty to animals)
	0	Otherwise
Crime Severity		A Bail Agency severity code for the gravity of the most serious charge as reflected in maximum and minimum sentences prescribed by law. Range: 1–135, where 1 = first-degree murder and 135 = soliciting for prostitution (Welsh and Viets, 1976, pp. D1–D8)

DEFENDANT'S RESOURCES

Variable	Code	Definition
Income		Defendant's self-reported legitimate income (0 = none or missing; 1 = \$0–80/week, 2 = \$80–100/week, 3 = \$100–120/week, 4 = \$120–160/week, 5 = \$160–200/week, 6 = \$200–300/week, 7 = \$300–360/week, 8 = more than \$360/week)
Income Unknown	1	If the income report is missing
	0	Otherwise
Education		Defendant's self-reported years of schooling (general education degree coded as 12; 17 = 17 or more)

Purging (Excluded) Exogenous Variables

Variable	Code	Definition
Concurrent Address	1	If defendant considers self to have more than one permanent address (e.g., parents' and girlfriend's homes)
	0	Otherwise
Time at Address		Defendant's self-reported months at current address (0 = no permanent address; 98 = eight years or more)
Time in D.C.		Self-reported number of years defendant has resided in D.C. area (0 = nonresident)
Non-D.C. Resident	1	If defendant is not a self-reported resident of D.C.
	0	Otherwise
Age		Defendant's self-reported age in years
Employed Off and On	1	If defendant reports he is employed on irregular basis
	0	Otherwise
Lives with Family	1	If defendant reports living with immediate or other family members
	0	Otherwise

Variable	Value	Definition
Mental Record	1	If defendant admits having been hospitalized in a mental hospital within previous ten years
	0	Otherwise
Alcohol	1	If defendant admits to being an alcoholic or having been treated for alcoholism
	0	Otherwise
Narcotics	1	If defendant admits prior or current drug addiction or treatment
	0	Otherwise
Bond Status	1	If defendant was on pretrial release for another crime when arrested
	0	Otherwise
Under Sentence	1	If defendant is on probation, parole, work release, or diversion from another sentence
	0	Otherwise
Prior Skips		Number of times defendant has failed to appear for previous court proceedings, excluding traffic and municipal violations (8 = 8 or more)
Felony	1	If most serious charge is a felony
	0	Otherwise
Morals Offense	1	If most serious charge relates to morals (dangerous drugs, gambling, sex offenses)
	0	Otherwise
Public Order Offense	1	If most serious charge relates to public order (weapons, obstruction of justice, bribery, flight and escape, parole/probation violation, rioting, possession of crime tools, harboring a fugitive, and introducing contraband into a penal institution)
	0	Otherwise

Variable definitions for Tables 2 and 3. The data come from the 1975 defendant survey by the Washington, D.C., Bail Agency (Welsh and Viets, 1976).

Since both the dependent variable (guilt) and the treatment variable (pretrial release) are dichotomous, neither ordinary probit analysis nor probit analysis with a purged treatment effect produces trustworthy (consistent) estimates. Instead the specification of the outcome equation was assumed to be a linear probability model with an endogenous treatment variable. The four-step G2SLS estimator discussed earlier in this chapter was used, in which ordinary two-stage least squares is first applied, followed by another two-stage estimation with heteroskedasticity corrected.

The validity of the linear probability specification was checked in three ways. First, forecasts were examined for out-of-bounds values (above 1.0 or below 0.0). None were found: the forecasts had a mean of 0.45 and a standard error of 0.08, with none above 0.70 or below 0.11. Second, forecasts were grouped into intervals and compared with actual proportions in the sample. As one would expect with forecasts covering a relatively small range, the fit was good and the sample showed no obvious signs of nonlinearity.

Third and most important, the coefficients themselves were examined for substantive meaning and manageable standard errors. Again no difficulties appeared: variables whose true direction is known a priori had effects in the expected direction. More income or education reduces the probability of conviction, for example, while additional prior convictions increase it (over the 0 to 5 range that includes nearly all defendants). Moreover, the marginal impact of an additional conviction lessens as their number increases. It is also noteworthy that women and blacks are somewhat (about 4 percent) more likely to be convicted than whites and males. Whether this is discrimination or simply a reflection of unmeasured differences cannot be determined without more information. Finally, though this is to some degree an artifact of the large sample and selection of controls for crime class, all control variables are significant beyond the 0.05 level and most are significant beyond 0.01.

Table 2 gives the substantive conclusion from the analysis, along with contrasting results from the two other methods used in previous work—simple comparison of conviction rates between released and detained defendants, and ordinary regression with the control variables entered. The three estimates of the deleterious impact of pretrial detention are as follows. The naive estimate based on simple compari-

TABLE 2. HOW MUCH DOES PRETRIAL DETENTION INCREASE THE PROBABILITY OF CONVICTION?

	Percentage Points (Standard Error)
Raw difference between released and detained groups	5.7** (1.2)
Difference with key variables controlled	4.1** (1.3)
Difference with key variables controlled and nonrandom assignment corrected	0.6 (3.8)

**Significant at 0.01 level.

Three estimates of the discriminatory effect of pretrial detention on the conviction rate. Data are taken from the 1975 District of Columbia Bail Agency sample (Welsh and Viets, 1976). For details, see Table 3.

sons of conviction rates is 6 percent and the regression estimate is 4 percent, both statistically significant beyond the 0.01 level. The simultaneous-equation estimate is just one-half of 1 percent, however, which is neither statistically nor substantively significant. In short, for this sample the best guess is that the apparent discriminatory effect of pretrial detention is purely artifactual. (The full results of the simultaneous-equation and regression analyses are given in Table 3.)

This conclusion was not altered under a variety of alternative specifications. To assess the adequacy of the exclusion restrictions, a stepwise regression was run; the excluded exogenous variables were the candidates for entry. Eighteen additional simultaneous-equation estimates of the treatment effect resulted. One of them showed a larger effect than the original estimate—three-fourths of 1 percent. The other seventeen estimates were all *smaller* than one-half of 1 percent, including eleven that were (very slightly) on the other side of zero. Perhaps the best summary of these specification tests is that they lend the most weight to the hypothesis that pretrial detention has no effect at all on conviction rates.

TABLE 3. PREDICTION EQUATIONS FOR PROBABILITY THAT A DEFENDANT IS CONVICTED—COEFFICIENTS (STANDARD ERRORS).

Variable	GLS	G2SLS
Released	−0.041** (0.013)	−0.0057 (0.0375)
Black	0.036* (0.016)	0.043** (0.017)
Male	−0.050** (0.014)	−0.045** (0.014)
Prior Convictions	0.031** (0.008)	0.033** (0.008)
(Prior Convictions)^2	−0.0031** (0.0012)	−0.0035** (0.0012)
Violent Offense	−0.089** (0.023)	−0.085** (0.024)
Dangerous Offense	0.053** (0.019)	0.043* (0.020)
Crime Against Person	−0.119** (0.019)	−0.121** (0.019)
Crime Against Property	−0.077** (0.011)	−0.081** (0.012)
Crime Severity	−0.0017* (0.0007)	−0.0020* (0.0008)
(Crime Severity)^2	0.000012** (0.000004)	0.000013** (0.000005)
Income	−0.0057* (0.0025)	−0.0075** (0.0026)
Income Unknown	−0.032* (0.015)	−0.036* (0.015)
(Education)^2	−0.00049** (0.00010)	−0.00050** (0.00010)
Intercept	0.656** (0.037)	0.634** (0.049)
N	11,044	10,365

* Significant (two-tailed) at 0.05.
** Significant (two-tailed) at 0.01.

Two linear equations for the probability that a defendant will be judged guilty. GLS: linear probability model corrected for heteroskedasicity. G2SLS: linear probability model with release treated as endogenous and heteroskedasicity corrected. Data are taken from the 1975 District of Columbia Bail Agency sample (Welsh and Viets, 1976). Sample sizes differ somewhat due to missing data on excluded exogenous variables (see Table 1).

If none of the credible evidence supports a discriminatory impact of pretrial detention, neither does it exclude the possibility entirely. The standard error of the properly estimated effect is nearly four percentage points. Although this value makes unlikely a true effect as large as 6 percent (less than 0.08 probability), a true effect of 4 percent lies well within the usual confidence intervals ($p = 0.19$). Thus modest discriminatory effects remain a possibility. What the analysis has shown is not that discriminatory effects of this size are excluded by the data, but rather that they are not favored. By contrast, the hypothesis of negligible effects, which seemed excluded in the inappropriate analyses, is now seen to be the most plausible of all.

Appendix

In the generalized least squares (GLS) problem in which $y = X\beta + u$, $E(u) = 0$, and $E(uu') = \Sigma$, the coefficient vector is usually estimated by $\hat{\beta} = (X'\hat{\Sigma}^{-1}X)^{-1}X'\hat{\Sigma}^{-1}y$, where $\hat{\Sigma}$ is an estimate of Σ. It is sometimes asserted in informal econometric studies that if each element of $\hat{\Sigma}$ is consistent for the corresponding element of Σ, then $\hat{\beta}$ is consistent. Sometimes this argument is made more explicit by assuming that $\hat{\Sigma}$ depends continuously on a fixed, finite number of consistent estimates of some parameter vector. For example, Σ may be diagonal with nonzero elements $(Z_i\gamma)^2$ $(i = 1, \ldots, n)$, where n is the number of observations, Z_i is a row vector of known constants, and γ is a parameter vector that is consistently estimated by $\hat{\gamma}$. The diagonal elements of Σ may then be estimated by $(Z_i\hat{\gamma})^2$. Then, it is said, $\hat{\Sigma}$ is a continuous function of a fixed, finite number of consistent estimates and therefore $\hat{\beta}$ is also consistent.

This argument fails in general. Slutzky's theorem does indeed guarantee that a continuous function of a finite number of consistently estimated parameters will converge in probability to the value of the function at the true parameter values. The difficulty in the GLS case is that the function is not fixed from observation to observation. As $n \to \infty$, the dimensions of Σ grow, altering the function that takes the parameter estimate used to estimate $\hat{\Sigma}$ into the coefficient estimate $\hat{\beta}$. Except in certain special circumstances, then, $\hat{\beta}$ is not necessarily consistent. Indeed it is not difficult to construct an example in which $\hat{\beta}$ would be consistent if Σ were known but in which it is not consistent if $\hat{\Sigma}$ is used instead, even though $\hat{\Sigma}$ depends continuously on a fixed, finite number of consistently estimated parameters.

Similarly, in the nonlinear simultaneous equation problem, it is sometimes asserted that if the reduced-form expectation of an endogenous variable (the purged value in the linear case) can be consistently estimated, then it can be inserted on the right-hand side of a structural equation and ordinary regression can then be applied (as in 2SLS). Consistent estimates are said to result. The implicit argument again depends on Slutzky's theorem, and it fails for the same reason as in the GLS case.

In this appendix, these imprecisions are corrected. Here we consider the case of a structural equation model that is linear in parameters but may be nonlinear in endogenous variables. Moreover, this equation may be heteroskedastic, so that both reduced-form expectations and the variance-covariance matrix of the disturbances must be estimated. Conditions are given which ensure that certain coefficient estimates are consistent, and their sampling variances are given. First, the lemmas give supplementary conditions under which the appeal to Slutzky's theorem is valid. All operations are carried out in euclidean space.

Lemma 1. Let A and B be $n \times k$ matrices of constant elements that are uniformly bounded in absolute value by m, and let Λ be an $n \times n$ diagonal matrix whose diagonal elements are equicontinuous functions of a k-dimensional stochastic vector $\hat{\gamma}$. Suppose that as $n \to \infty$, $\hat{\gamma}$ converges in probability to γ and that $\Lambda = 0$ at $\hat{\gamma} = \gamma$. Finally, let u be an n-dimensional vector of independent observations on random variables, each of which has mean zero and variance σ^2. Suppose that the latter variables meet the Lindeberg–Feller conditions (Dhrymes, 1970, p. 106) or are identically distributed. Then plim $A'\Lambda B/n$ = plim $A'\Lambda u/\sqrt{n} = 0$.

Proof: Denote the diagonal elements of Λ by λ_k and the typical element of $A'\Lambda B/n$ by g_{ij}. By the equicontinuity assumption, for sufficiently large n every λ_k is within a given ϵ of zero with arbitrarily high probability. Hence:

$$\begin{aligned} \text{plim } |g_{ij}| &= \text{plim} \left| \frac{\sum_{k=1}^{n} a_{ik}\lambda_k b_{kj}}{n} \right| \\ &\leq \frac{\lim m^2 \sum \epsilon}{n} \\ &= \lim m^2 \epsilon \end{aligned}$$

which may be made arbitrarily small.

For the second result, denote the typical element of $A'\Lambda u/\sqrt{n}$ by h_{ij}. Thus:

$$h_{ij} = \frac{\sum a_{ik}\lambda_k u_k}{\sqrt{n}}$$

Then with arbitrarily high probability h_{ij} has a density function bounded above by that of the random variable $m\epsilon\sum u_k/\sqrt{n}$.

Since the u_k meet the Lindeberg–Feller conditions (or else are i.i.d.), the latter sum converges to a centered normal distribution with variance σ^2 (Dhrymes, 1979, pp. 104–105). But since plim $m\epsilon$ may be chosen arbitrarily small, it follows that plim $h_{ij} = 0$.

Lemma 2. Under the conditions of Lemma 1, suppose that each element of each column j of $\hat{A}$ is an equicontinuous function of a finite-dimensional parameter vector $\hat{\alpha}_j$ that converges in probability to α_j and that when $\hat{\alpha}_j = \alpha_j$ for all j, then $\hat{A} = A$. Similarly, the elements of each column j of $\hat{B}$ are equicontinuous functions of $\hat{\beta}_j$, $\hat{\beta}_j$ is consistent for β_j, and $\hat{\beta}_j = \beta_j$ for all j implies $\hat{B} = B$. Let D be an $n \times n$ diagonal matrix of constants bounded in absolute value by M such that lim $A'DB/n = G$, a fixed matrix. Then plim $\hat{A}'\Lambda\hat{B}/n =$ plim $\hat{A}'\Lambda u/\sqrt{n} = 0$ and plim $\hat{A}'D\hat{B}/n = G$.

Proof: Since the elements of A and B are bounded in absolute value by m, and since each column of A and B is subject to equicontinuity conditions and the number of such columns is fixed, then for arbitrary $\epsilon > 0$ it follows that each element of $\hat{A}$ and $\hat{B}$ is bounded in absolute value by $m + \epsilon$ with arbitrarily high probability. The proof of the first two results then proceeds as in Lemma 1.

For the last claim, note that:

$$\hat{A}'D\hat{B} = A'DB + (\hat{A} - A)'DB + A'D(\hat{B} - B) + (\hat{A} - A)'D(\hat{B} - B)$$

The last three terms on the right have zero probability limits when divided by n. For example, denoting the diagonal elements of D by d_k and the typical element of $(\hat{A} - A)'DB/n$ by g_{ij}, then, as in Lemma 1:

$$\begin{aligned} \text{plim}|g_{ij}| &\le \text{plim } m\epsilon \sum \left|\frac{d_k}{n}\right| \\ &\le \lim mM\epsilon \end{aligned}$$

Thus plim$|g_{ij}| = 0$. These results then imply plim $\hat{A}D'\hat{B}/n =$ lim $AD'B/n$.

In these first two lemmas, the least intuitive requirement is the equicontinuity of the functions used to estimate the matrices. In circumstances frequently encountered in econometrics, however, this condition is implied by relatively mild restrictions on the functions and their arguments. The next lemma essentially considers the case in which each matrix element to be estimated is computed as the same function of a forecast value from some equation. The forecasts might be squared, for example. Thus while the forecasts vary from observation to observation, the function that relates these forecasts to the estimates of the matrix elements is fixed. In this case, the next lemma shows that boundedness and continuity conditions suffice to ensure equicontinuity of the functions used to estimate the matrix elements.

Lemma 3. Let $f(\cdot)$ be a continuous function from the closed interval I to the real line. Let γ be a g-dimensional (column) vector, and for each k $(k = 1, \ldots, n)$ let Z_k be a g-dimensional (row) vector whose elements are uniformly bounded in absolute value by m. Finally, suppose that the scalar $Z_k\gamma$ falls in I for all k. Then the functions $h_k(\gamma) = f(Z_k\gamma)$ are equicontinuous functions of γ and are uniformly bounded.

Proof: It must be shown that for arbitrary $\epsilon > 0$ there exists δ that does not depend on γ_0 or k, such that $\| \gamma - \gamma_0 \| < \delta$ implies $| f(Z_k\gamma) - f(Z_k\gamma_0) | < \epsilon$. Now f is a continuous function over a closed, bounded interval and hence is uniformly continuous. Moreover, f is the same function for all k. Hence there exists a single δ^* such that, for all γ_0 and k,

$$|Z_k\gamma - Z_k\gamma_0| < \delta^* \qquad \text{implies} \qquad | f(Z_k\gamma) - f(Z_k\gamma_0) | < \epsilon$$

Since the absolute values of the elements of X_k are bounded by m for all k, however, it follows that, for all k and γ_0,

$$\| \gamma - \gamma_0 \| < \delta \qquad \text{implies} \qquad |Z_k\gamma - Z_k\gamma_0| < m\delta$$

Thus if δ is chosen so that $m\delta < \delta^*$, $| f(Z_k\gamma) - f(Z_k\gamma_0)| < \epsilon$ for all γ_0 and k, as required. Uniform boundedness then follows from the continuity of f over a compact interval.

The main result now follows. Let:

$$\begin{aligned} y_1 &= G(Y, X)\,\gamma + X_1\beta + u \\ &= H(Y, X)\delta + u \end{aligned} \tag{A-1}$$

be the first structural equation in a system, where y_1 is an $n \times 1$ vector of observations on the first endogenous variable, $G(Y, X)$ is a matrix of observations on m_1 possibly nonlinear functions of the m endogenous variables Y and the k exogenous variables X in the system, X_1 is the $n \times k_1$ matrix of observations on the included exogenous variables, u is the $n \times 1$ vector of disturbances, and γ and β are m_1 and k_1-dimensional parameter vectors to be estimated. Finally, $H = [G(Y, X), X_1]$ and $\delta' = [\gamma', \beta']$; and δ is identified by exclusion restrictions in Equation (A-1).

It is assumed that:

(i) $E(u|X) = 0$, $E(uu'|X) = \Sigma$, where Σ is an unknown diagonal matrix with positive diagonal elements. The elements of u meet the Lindeberg–Feller conditions.

(ii) There is a g-dimensional column vector δ and a set of g-dimensional row vectors Z_i $(i = 1, \ldots, n)$ whose elements are bounded in absolute value such that each diagonal element of Σ^{-1} can be written as $f(Z_i\delta)$, where f is a continuous function from some closed interval I to the real line, and such that $Z_i\delta$ is in I for all i. A consistent estimate of δ exists, and the implicit estimator of Σ^{-1} is denoted $\hat{\Sigma}^{-1}$.

(iii) $E[G(Y, X)|X] = \Pi(X)$ exists and is not collinear with X_1. This matrix will be called "the expected value of the reduced form."

(iv) For each column j of $\Pi(X)$ $(j = 1, \ldots, m_1)$, there is a p_j-dimensional column vector γ_j and a set of p_j-dimensional row vectors W_i^j that are functions of X and whose elements are bounded in absolute value. Each element i of the jth column of $\Pi(X)$ can be written as $h_j(W_i^j\gamma_j)$, where h_j is a continuous function from the closed interval I_j to the real line, such that $W_i^j\gamma_j$ belongs to I_j for all i, j. A consistent estimate of each γ_j exists, and the implicit estimator of Π is denoted $\hat{\Pi}$.

(v) Set $P = [\Pi(X), X_1]$. Then $P'P/n$ and $P'\Sigma^{-1}P/n$ converge to fixed positive definite matrices.

Assumptions (i) and (v) imply:

(vi) $\text{plim}\, P'u/n = \text{plim}\, P'\Sigma^{-1}u/n = 0$.

The estimator $\hat{\delta}_{LS}$ is defined by the following procedure. First the consistent estimator of the expectation of the reduced form, $\hat{\Pi}$, is adjoined to X_1 to form the matrix of instruments, $\hat{P}$. The entire right-hand side of the original structural equation is then purged with these instruments in the usual way, giving the matrix of purged values, $\hat{H}$. Then y_1 is regressed on $\hat{H}$. (An equivalent but computationally easier method is discussed in the main body of this chapter and Chapter 6.)

Formally, denote by $\hat{P}$ the matrix of instruments $[\hat{\Pi} \quad X_1]$, and let $\hat{H} = \hat{P}\,(\hat{P}'\hat{P})^{-1}\hat{P}'H$ be the purged values of H. To see that this estimator very likely exists in large samples, note that Lemma 2, with D set equal to the identity matrix, implies that $\hat{P}'\hat{P}/n$ converges in probability to $\lim P'P/n$, whose inverse exists. But then the inverse of $\hat{P}'\hat{P}/n$ also exists with arbitrarily large probability, since the determinant is a continuous function of the finite number of elements of $\hat{P}'\hat{P}/n$ and hence is bounded away from zero in probability.

Now $\hat{H}$ is the matrix of predicted values of the reduced form of H, namely P. Since the matrix of regressors is a consistent estimate of P itself, the lemmas show that the elements of $\hat{H}$ converge to P. That is, $\hat{H}$ is another estimate of P.

Define the estimator:

$$\hat{\delta}_{LS} = (\hat{H}'\hat{H})^{-1}\hat{H}'y_1 \tag{A-2}$$

By the same argument given above, this estimator exists with arbitrarily high probability in large samples. The lemmas show that $\text{plim}\ \hat{H}'\hat{H}/n = \lim P'P/n$ and $\text{plim}\ \hat{H}'u/n = \text{plim}\ P'u/n = 0$. Moreover, $\hat{H}'\,(H - \hat{H})/n = 0$ by the standard result that OLS forecasts and residuals are uncorrelated. Hence from (A-1) and (A-2):

$$\begin{aligned} \hat{\delta}_{LS} &= \frac{n(\hat{H}'\hat{H})^{-1}\hat{H}'[\hat{H}\delta + (H - \hat{H})\,\delta + u]}{n} \\ &= \delta + \frac{n(\hat{H}'\hat{H})^{-1}\hat{H}'u}{n} \end{aligned} \tag{A-3}$$

Thus $\text{plim}\ \hat{\delta}_{LS} = \delta$, and the estimator is consistent.

The asymptotic variance of $\sqrt{n}\hat{\delta}_{LS}$ is defined as the variance of the estimator to which $\sqrt{n}\hat{\delta}_{LS}$ converges in probability.[4] The lemmas show that $\sqrt{n}\hat{\delta}_{LS}$ converges to $n(P'P)^{-1}P'u/\sqrt{n}$, so that:

$$\text{var}\,(\hat{\delta}_{LS}) = (P'P)^{-1}P'\Sigma P(P'P)^{-1} \tag{A-4}$$

---

[4]The conditions on the disturbances and on the elements of P imply that $\hat{\delta}_{LS}$ converges in distribution to a normally distributed variable (Theil, 1971, pp. 479–99). A similar argument applies to $\hat{\delta}_{GLS}$.

In the homoskedastic case, $\Sigma = \sigma^2 I$ and (A-4) reduces to:

$$\text{var}(\hat{\delta}_{LS}) = \sigma^2(P'P)^{-1}$$

The latter estimator is a specialization of an estimator due to Jorgenson and Laffont (1974). If it is further specialized to require that $\Pi(X)$ be linear in X so that $\hat{P} = X(X'X)^{-1}X'H$, then $\hat{H} = \hat{P}$. In this case, $\hat{\delta}_{LS}$ reduces to the usual two-stage least squares estimator.

Returning now to the heteroskedastic case, suppose for concreteness that y_1 is dichotomous, so that if σ_i^2 is the ith diagonal element of Σ, then $\sigma_i^2 = H_i\delta(1 - H_i\delta)$, where H_i is the ith row of H. A consistent estimate $\tilde{\sigma}_i^2$ is $H_i\hat{\delta}_{LS}(1 - H_i\hat{\delta}_{LS})$. In this case, one might assume that for some $\rho > 0$, $\sigma_i^2 \geq \rho$ for all i. If one then defines $\hat{\sigma}_i^2 = \max(\tilde{\sigma}_i^2, \rho)$, this estimator is also consistent and meets the conditions of Assumption (ii) if H is bounded.

For any heteroskedastic setup meeting the assumptions, let $\hat{\Sigma}$ be the matrix Σ with $\hat{\sigma}_i^2$ substituted for σ_i^2, and define $\hat{\Sigma}^{-1/2}$ in the obvious way. Set $\hat{P}^+ = \hat{\Sigma}^{-1/2}\hat{P}$, $H^+ = \hat{\Sigma}^{-1/2}H$, and $y_1^+ = \hat{\Sigma}^{-1/2}y_1$. Let:

$$\hat{H}^+ = \hat{P}^+(\hat{P}^{+\prime}\hat{P}^+)^{-1}\,\hat{P}^{+\prime}H^+ = \hat{\Sigma}^{-1/2}\hat{P}(\hat{P}'\hat{\Sigma}^{-1}\hat{P})^{-1}\hat{P}'\hat{\Sigma}^{-1}H$$

As before, this estimator exists in large samples; by the lemmas, plim $\hat{H}^{+\prime}\hat{H}^+/n$ converges to $P^{+\prime}P^+/n = P'\Sigma^{-1}P/n$; and $\hat{H}^{+\prime}(H^+ - \hat{H}^+)/n = 0$.

We now define the generalized version of $\hat{\delta}_{LS}$. This estimator consists of applying $\hat{\delta}_{LS}$ first, using those coefficients to estimate Σ, then correcting for heteroskedasticity and applying $\hat{\delta}_{LS}$ again. Formally define:

$$\hat{\delta}_{GLS} = (\hat{H}^{+\prime}\hat{H}^+)^{-1}\hat{H}^{+\prime}y_1^+ \tag{A-5}$$

$$\begin{aligned} &= (\hat{H}^{+\prime}\hat{H}^+)^{-1}\,\hat{H}^{+\prime}[\hat{H}^+\delta + (\hat{H}^+ - \hat{H}^+)\delta + \hat{\Sigma}^{-1/2}u] \\ &= \delta + \frac{n(\hat{H}^{+\prime}\hat{H}^+)^{-1}\,\hat{H}^{+\prime}\hat{\Sigma}^{-1/2}u}{n} \end{aligned} \tag{A-6}$$

This estimator exists with arbitrarily high probability in large samples, and plim $\hat{H}^{+\prime}\hat{\Sigma}^{-1/2}u/n = \text{plim}\; P'\Sigma^{-1}u/n = 0$, so that it is consistent.

From Equation (A-6), $\sqrt{n}\hat{\delta}_{GLS}$ converges in distribution to $\delta + n(P^{+\prime}P^+)^{-1}P^+\Sigma^{-1/2}u/\sqrt{n}$, so that:

$$\text{var}(\hat{\delta}_{GLS}) = (P^{+\prime}P^+)^{-1} \tag{A-7}$$

The variance in (A-7) may be estimated consistently by $(\hat{H}^{+\prime}\hat{H}^+)^{-1}$. The usual least-squares calculations applied to (A-5) will give an estimate of var ($\hat{\delta}_{GLS}$), namely $\hat{\omega}^2(\hat{H}^{+\prime}\hat{H}^{+})^{-1}$, where $\hat{\omega}^2$ is the sample variance of the residuals $(H^{+} - \hat{H}^{+})\,\delta + \hat{\Sigma}^{-1/2}u$. Hence the correct variances can be obtained by correcting the estimate of the disturbance variance (which in this case is known to be 1), just as with 2SLS calculations.

To show that $\hat{\delta}_{GLS}$ is asymptotically best in the class of consistent single-equation instrumental variable (IV) estimators, let an alternative estimator in this class be:

$$\tilde{\delta} = (D'H)^{-1}\,D'y_1$$
$$= \frac{B'\Sigma^{-1}y_1}{n}$$

where D is an arbitrary $n \times (m_1 + k_1)$ matrix and $B' = (D'H/n)^{-1}D'\Sigma$. To make $\tilde{\delta}$ consistent with an asymptotic variance, it is assumed that plim $D'H/n$ and lim $D'\Sigma D/n$ are constant nonsingular matrices and that, conditional on D, u and H have the same properties as in Assumption (i) above. Then asymptotically:[5]

$$\text{plim}\,\frac{D'u}{n} = 0 \qquad \text{(A-8)}$$
$$E(D'H) = D'P$$

From (A-5), $\hat{\delta}_{GLS}$ has the same asymptotic distribution as:

$$\bar{\delta} = \frac{C'\Sigma^{-1}y_1}{n}$$

where $C' = (F'\,\Sigma^{-1}H/n)^{-1}F'$ and $F' = H'\Sigma^{-1}P(P'\Sigma^{-1}P)^{-1}P'$. Thus $\bar{\delta}$ is also an IV estimator. Note that:

$$\frac{B'\Sigma^{-1}H}{n} = \frac{C'\Sigma^{-1}H}{n} = I \qquad \text{(A-9)}$$

and hence asymptotically by (A-8):

$$E\left(\frac{B'\Sigma^{-1}H}{n}\right) = E\left(\frac{B'\Sigma^{-1}P}{n}\right) = E\left(\frac{C'\Sigma^{-1}P}{n}\right) \qquad \text{(A-10)}$$

---

[5]To allow for the possibility that the elements of the matrix of instruments are random in finite samples but converge in probability to constants, the asymptotic expectation of a random variable x refers to the expectation of the random variable to which x converges in probability.

Now let:

$$\sqrt{n}\tilde{\delta} = \frac{C'\Sigma^{-1}y_1}{\sqrt{n}} + \frac{(B-C)'\Sigma^{-1}y_1}{\sqrt{n}}$$
$$= \sqrt{n}\,\bar{\delta} + \frac{(B-C)'\Sigma^{-1}u}{\sqrt{n}} + \frac{(B-C)'\Sigma^{-1}H\delta}{\sqrt{n}}$$

By (A-9), the last term is zero, so that asymptotically:

$$\text{var}\,(\sqrt{n}\tilde{\delta}) = \text{var}\,(\sqrt{n}\,\bar{\delta}) + E\left[\frac{(B-C)'\Sigma^{-1}uu'\Sigma^{-1}(B-C)}{n}\right] + E(A + A')$$

where $A = (B-C)'\Sigma^{-1}uu'\Sigma^{-1}C/n = (B-C)'\Sigma^{-1}PR/n$ and where R is a matrix that converges to a constant. Hence by Equation (A-10), $E(A) = 0$. Then

$$\text{var}\,(\tilde{\delta}) = \text{var}\,(\hat{\delta}_{GLS}) + \frac{[\text{plim}(B-C)'\Sigma^{-1}(B-C)/n]}{n}$$

which exceeds var $(\hat{\delta}_{GLS})$ by a positive semidefinite matrix.

This method of proof is due to Basmann (1957), who used it to show that in the linear homoskedastic case 2SLS is "best linear consistent" within the class of IV estimators. The following result has been established:

Theorem. Under Assumptions (i) to (v), let $y_1 = H(Y,X)\delta + u$ as in (A- 1). Then the estimator $\hat{\delta}_{LS}$ defined in (A- 2) is consistent and has asymptotic variance given by (A-4). The estimator $\hat{\delta}_{GLS}$ defined in (A- 5) is also consistent and has asymptotic variance given by (A-7). Moreover, the asymptotic variance of any other consistent single-equation IV estimator exceeds that of $\hat{\delta}_{GLS}$ by a positive semidefinite matrix.

4

Quasi-Experiments with Censored Data: Why Regression and Weighting Fail

CENSORED DATA

Some quasi-experiments yield data only from a subsample of the population of interest. As noted in the first chapter, research designs of this type appear throughout the social sciences. For example, survey researchers aim for complete coverage of a population but typically achieve much less. Some respondents cannot be located and others refuse, skewing the sample toward older, more stable, more rural respondents. The problem is particularly acute in panel studies, in which respondents must be interviewed several times over periods of months or years. Response rates frequently drop to 50 percent and below, meaning that a nonrandom *censored* sample of interviewees must be used to draw inferences about the full population. (See, for example, Kish, 1965, chap. 13.)

The same problem occurs in evaluation research. Suppose that the admissions committee for a university graduate program wishes to know how much college grades should be weighed in assessing applicants. Since the criterion is "success in the graduate program," statistical evidence will be available only from previous classes—the group of applicants who actually enrolled. Information about performance is thus restricted to a group chosen for its likely success in the program. The admissions committee wishes to forecast outcomes in the pool of applicants, but it is forced to rely solely on experience with a nonrandom subset of them. (Interesting studies in this field include Nicholson, 1970; Dawes, 1975; Astin, 1977.)

At first glance, censored samples like these appear to require no special statistical methods. In the admission problem, one may collect a sample of graduate students enrolled in the program in previous years, measure their performance (graduate school grades, professional prominence after the doctorate, or some other index), and regress this measure on college grade averages. The resulting coefficient would be taken as an indicator of the predictive power of college grades. If desired, additional variables could be entered as controls. Since there is no simultaneity here (grades are not caused by subsequent success), there seems to be no reason to employ any but ordinary regression techniques. Apparently the nonrandom sample should be treated as if it were random. This practice—ignoring the nature of the sample—is virtually universal.

Elementary statistical methods break down, however, when confronted by censored samples. To see why, consider the following simplified statistical model of the graduate admissions process. (As in Chapter 2, capital letters denote the name of a variable and lowercase letters denote what the variable measures.)

Selection equation:

$$\text{Admission Rating} = a_1 + b_1\,(\text{Grades}) + u_1 \tag{1a}$$

$$\text{Admission} = \begin{cases} 1 & \text{if Admission Rating} \geqslant 0 \\ 0 & \text{if Admission Rating} < 0 \end{cases} \tag{1b}$$

Outcome equation:

$$\text{Success} = \begin{cases} a_2 + b_2\,(\text{Grades}) + u_2 & \text{if Admission} = 1 \\ \text{unobserved} & \text{if Admission} = 0 \end{cases} \tag{2}$$

This model asserts that admission to graduate school occurs when an applicant's Admission Rating is sufficiently high. If Admission Rating exceeds zero, the applicant is accepted and Admission = 1; if Admission Rating is below zero, the applicant is not admitted and Admission = 0. Admission Rating is a hypothetical measure of the overall strength of a candidate for admission and it is not observed. However, this evaluation is known to be based on college Grades (expressed as a grade point average, for instance) plus the net effect of all unmeasured factors—namely the disturbance u_1, which is assumed to be uncorrelated with Grades. This residual quantity might incorporate the effects of age, character strength, and other unmeasured factors influencing the admissions committee's decisions; it is postulated to be normally distributed with mean 0 and variance 1. Together equations (1a) and (1b) constitute a probit specification for the admission decision; they constitute the selection mechanism.

In Equation (2), the dependent variable, Success (in graduate work), is observed only after the student has been admitted. It is assumed to be a linear function of Grades plus residual factors uncorrelated with Grades. This is the outcome equation.

In the conventional approach to this system of equations, the selection mechanism (1a) and (1b) would be ignored. Regression analysis would be applied to the outcome equation (2) as if the data were randomly sampled from the population of applicants. If a graduate student entered the program with low college grades, for instance, he or she would be regarded as representative of the pool of applicants with low grades.

Admitted graduate students are not representative of applicants generally, however, as the admission equations make clear. There are many college graduates with low grades who attempt to enroll in a graduate program; only a few succeed. These happy exceptions usually owe their success to favorable personal characteristics other than grades. In some instances, these additional factors may themselves be unrelated to subsequent success, as when students from distant areas are accepted to maintain a balanced student population. Such students are generally expected to do neither better nor worse than any other applicant with the same college grades, and in that sense they are statistically representative of the applicant pool. More commonly, however, an applicant's margin of superiority is academic in character, as when it is owed to evidence of maturity and determination. In that case the student is expected to perform above the level of other applicants with the same college grades, and he is no longer representative.

Now suppose that an admissions committee examines graduate student records to compare the performance of those who entered with low undergraduate grades to those with high undergraduate grades. The group of admittees with strong college transcripts will be approximately representative of that group of applicants, since nearly all such applicants are admitted and most will do well. However, weak college performers will be represented by just those members the admissions committee thought would perform best. If the committee members have any aptitude for their jobs, both the high-grades and the low-grades applicants will do reasonably well on average. The poorer college performers may well match the students with high grades, and they will certainly perform much better than typical students with poor college records. In the admitted group, students who entered with low grades may be as successful as those with good grades.

The committee might then be tempted to conclude that college grades do not predict success. Intuition makes clear, however, that this result does not extend to the *applicant group*, where students with low grades in general would perform quite poorly in graduate work. If a random sample of applicants were admitted, college grades would predict their success very well. Thus the committee is right to use grades as a discriminant in this example. Failure to predict success among *admitted* applicants is no evidence against any admission criterion. Discrimination matters only in the applicant pool, where grades may perform very well as a predictor. (Conclusions of this kind are demonstrated formally in some important special cases by Goldberger, 1972.)

HOW REGRESSION FAILS

Although many researchers are vaguely aware that censored samples create special challenges to inference, the nature of those challenges is poorly understood. For deeper insight into the bias introduced when ordinary regression is applied to a censored sample, a more formal treatment is required. In the model defined by Equations (1) and (2), assume that Grades and the two disturbance terms are normally distributed, thereby ensuring that Admission Rating and Success (which are linear functions of Grades and the disturbances) are also normally distributed. Under these assumptions, when least squares is applied to the outcome equation (2) as though the sample were random, inconsistency results. Specifically, the estimated Grades coefficient converges asymptotically to a position between its true value b_2 and a certain other quantity denoted by b_2'.

The number b_2' is defined in the following way. Suppose that Success were observed for everyone in the population and that an "auxiliary regression" were estimated on that population:

Auxiliary equation:

$$\text{Success} = a_2' + b_2' \text{ (Grades)} + b_3 \text{ (Admission Rating)} + u_2' \qquad (2')$$

Notice that this equation is just the true outcome equation with the variable Admission Rating added as an explanatory factor. The coefficients in this equation are then defined by the values to which they would converge if this equation were estimated by OLS in the full population. In particular, b_2' is the coefficient that the variable Grades takes on in this equation. (Of course, it is impossible to carry out this regression in practice. Not only is Admission Rating not observed for any applicant, but Success is also missing for applicants who did not enroll. The equation is solely a theoretical tool.)

The nature of the asymptotic bias when OLS is applied to the censored sample of outcomes may now be stated precisely. Let the censored-sample OLS estimate of b_2 be denoted by $\hat{b}_2$ and let plim($\hat{b}_2$) be the value to which $\hat{b}_2$ converges asymptotically (in probability). Then for some λ, $0 < \lambda < 1$:

$$\text{plim}(\hat{b}_2) = \lambda b_2 + (1 - \lambda)\, b_2' \qquad (3)$$

Moreover, as the proportion of applicants selected becomes smaller, λ tends to 0. Thus as the selection grows more stringent, plim($\hat{b}_2$) will tend toward b_2'.

An informal inspection of Equation (2′) helps expose the meaning of the term b_2'. That equation is "misspecified" due to the presence of the spurious variable Admission Rating, which has no causal effect of its own on Success. Admission Rating does have some predictive power, however, since it is partly a function of the unmeasured factors influencing the admission decision. (See Equation (1).) In general, at least some of these unmeasured factors influence Success. Since these factors are not controlled elsewhere in the equation, their impact is attributed to Admission Rating, giving it a positive coefficient. But since the variable Admission Rating is a function of both the latter factors and grades, giving it a positive coefficient allows Admission

Rating to "explain" not just the impact of the unmeasured factors but also part of the effect of grades. Just a fraction of the true impact of grades will remain unaccounted for. This truncated effect will be attributed to the variable Grades, thereby underestimating its coefficient. Hence the coefficient of Grades, b_2', will be below the true value b_2.

With these observations in mind, return to the OLS estimate of b_2 in the censored sample of outcomes. As noted above, the estimate b_2 will compromise between the true b_2 and b_2'. But b_2' has just been shown to be an underestimate of b_2. Hence the estimate $\hat{b}_2$ will be driven downward. Under certain conditions, the downward impact can be so powerful that a positive coefficient will appear negative, meaning that not even the signs of coefficients can be trusted in a censored sample. In any case, Grades will appear less powerful in predicting Success than it actually is.

Erroneous results from censored samples are not inevitable. Suppose, for example, that the unmeasured factors employed by the committee in its admission decisions are uncorrelated with the unmeasured factors that influence success—that is, suppose that the disturbances u_1 and u_2 are uncorrelated. This means that the committee members have available just one piece of information that actually predicts later performance, namely Grades. Nothing else they might know or intuit about applicants predicts performance. Under these assumptions, the zero-order relationship of Grades to Success is to be assessed.

Once again, ordinary regression is applied to the censored sample. Then, just as before, the estimated Grades coefficient $\hat{b}_2$ will tend to a weighted average of the true value b_2 and the quantity b_2':

$$\text{plim}(\hat{b}_2) = \lambda b_2 + (1 - \lambda)\, b_2' \qquad (3)$$

Here b_2' is defined in the same way as before: the value that the coefficient of Grades would take on if Equation (2′) were applied to a hypothetical full sample of observations:

$$\text{Success} = a_2' + b_2'\,(\text{Grades}) + b_3\,(\text{Admission Rating}) + u_2' \qquad (2')$$

We have seen that if u_1 and u_2 are positively correlated, $\hat{b}_2'$ tends to be an underestimate of b_2. When the two disturbances are uncorrelated, however, a different logic applies. In equation (2′), the independent variables are Admission Rating and Grades. But Admission Rating is a function only of Grades and u_1, and the Grades variable is controlled. Hence the spurious variable Admission Rating can acquire a nonzero coefficient only if u_1 adds to the explanatory power of the regression. However, u_1 is uncorrelated with u_2, which includes all the true explanatory factors excluded from this equation. (See Equation (2).) Thus u_1 explains no additional variance, and asymptotically the coefficient of Admission Rating will be zero. Hence no inconsistency is introduced by adding this variable to the regression, and the coefficients b_2 and b_2' are identical. Under these conditions, ordinary least squares applied to the censored sample is consistent. That is, $\text{plim}(\hat{b}_2) = \lambda b_2 + (1 - \lambda)b_2' = b_2$.

This result justifies stratified sampling with unequal sampling probabilities. Survey researchers frequently oversample certain subgroups in a population, randomizing selection of respondents within each group. In the framework set out above, selection to a stratified sample occurs as a function of group membership plus a purely random disturbance term, which is certainly uncorrelated with the disturbance in any other regression. Group memberships are then controlled in subsequent statistical analyses. Under these conditions, u_1 and u_2 are uncorrelated and no error is introduced by treating the stratified sample as though it were a simple random sample of the population. Correct estimates of statistical effects will follow.

In general, if *every* variable influencing selection is controlled in the outcome equation, the correlation between the disturbances disappears. In stratified sampling, these variables have been chosen by the researcher, and including them in the outcome equation presents no difficulties. Consistent coefficient estimates are then obtained by applying OLS to the censored sample. But of course most censored samples are obtained by complex selection processes, and the complete list of variables influencing selection is not measured or even known. Including them in the outcome equation is impossible, and biased coefficient estimates result.

In summary, the application of ordinary regression to a censored sample does not generally yield consistent coefficient estimates. In the cases considered here, where the true selection and outcome equations have only one independent variable each, it was shown that consistent estimates are obtained if and only if the disturbances in the selection and outcome equations are uncorrelated with each other. Thus if σ_{12} is the covariance between u_1 and u_2, consistent estimates in the outcome equation require that $\sigma_{12} = 0$. The appendix to this chapter shows that this result holds whenever the independent variables and disturbances are normally distributed, no matter how many such variables enter the two equations. In a censored sample, dependable coefficients are obtained by OLS if and only if unobserved factors influencing selection are uncorrelated with unobserved factors influencing outcomes.

More precisely, consider a censored-sample regression problem in which disturbances and all variables (except the intercept) are normally distributed. Observations have been generated by a probit specification like that in equations (1a) and (1b). Suppose further that the selection and outcome equations are properly specified. Then, in general, ordinary regression applied to the outcome equation in a censored sample is consistent if and only if the resulting disturbances from the two equations are uncorrelated ($\sigma_{12} = 0$).[1] Censored samples meeting this condition are said to be consistently censored.

In practice, censored samples rarely meet this requirement, and therefore inconsistent estimates are the norm. Typically, unobserved factors influencing selection also influence performance, inducing a correlation between u_1 and u_2. Ordinary regression then produces undesirable estimates, even if the researcher controls for all the measured factors that influence selection. Without measuring every variable, one cannot be sure that the unmeasured variables have not destroyed the validity of the results. In general, unless every factor that influences selection has been measured and controlled in the outcome equation, OLS does not adjust for the effects of censored sampling.

The precise nature of the inconsistency of the coefficients is more difficult to describe in the multiple regression case. But just as in bivariate regression, as the censoring becomes severe the coefficients eventually tend to those in an auxiliary regression in which Admission Rating is added to the outcome equation. Unlike the bivariate case,

---

[1]If the conditions are strengthened to require statistical independence of u_1 and u_2 and of x_1 and x_2, then the result holds for variables with nonnormal distributions.

however, for moderate amounts of censoring the coefficients may resemble neither their value in this misspecified regression nor their true values. That is, individual coefficients need not converge to values between those they would take on with no censoring or complete censoring. Nevertheless, the multiple regression case is not fundamentally different from the simpler cases considered above. There is a transformation of the independent variables such that for any amount of censoring, individual coefficients will fall between their true value and their value in the auxiliary equation. Moreover, as the censoring becomes heavier, the coefficients will tend to their values in the auxiliary equation. Thus with an appropriate transformation, each coefficient in the multiple regression behaves just as the single coefficient does in the bivariate case. (See the appendix.)

EVALUATING SELECTION PROCEDURES WITH OLS

When applicants are chosen for admission to an academic program, researchers commonly evaluate the caliber of the selection process by regressing subsequent performance by applicants on predictor variables such as prior grades and test scores. The argument is that the admissions committee is known to have used these predictors; if the variables fail to correlate with performance, the committee is thought to have erred. In the same way, when judges release certain arrestees before trial, the defendant's arrest record and the nature of the criminal charge are used to predict bail jumping or rearrests in the released groups. If the correlation is low, behavior is said to be unpredictable and the action of the judges arbitrary (Kirby, 1977).

To evaluate this logic, return to the admissions committee example in which the predictive power of Grades was to be assessed. Suppose that the committee evaluates Grades properly. That is, the admissions rating in equation (1a) is based on the same coefficients as the true success equation in (2) ($a_2 = a_1$ and $b_2 = b_1$), and the variance of the unobserved factors is identical. Furthermore, suppose that the committee is able to exploit some of the information in the unmeasured factors, so that admissions decisions are better than the measured variables alone could make them, giving u_1 and u_2 a correlation r. Since the variances are equal, r is also the regression coefficient of u_2

on u_1. Hence $u_2 = ru_1 + v_2$, where v_2 is a random error term. Thus equation (2) can be written as:

$$\begin{aligned} \text{Success} &= a_2 + b_2 \text{ (Grades)} + u_2 \\ &= a_1 + b_1 \text{ (Grades)} + ru_1 + v_2 \end{aligned} \quad (4)$$

Now if OLS is applied to the selected group, the estimated coefficient for Grades will fall between its true value in equation (4) and the value b_2' defined by equation (2′) as before. To find b_2' in this case, note that the variable Admission Rating appearing in (2′) is a function solely of Grades and u_1:

$$\text{Admission Rating} = a_1 + b_1 \text{ (Grades)} + u_1 \quad (1a)$$

Hence from equation (2′):

$$\begin{aligned} \text{Success} &= a_2' + b_2' \text{ (Grades)} + b_3 \text{ (Admission Rating)} + u_2' \\ &= a_2' + b_2' \text{ (Grades)} + b_3[a_1 + b_1 \text{ (Grades)} + u_1] + u_2' \\ &= (a_2' + b_3 a_1) + (b_2' + b_3 b_1) \text{ (Grades)} + b_3 u_1 + u_2' \end{aligned}$$

Since this last equation relates the same variables in the identical functional form as equation (4), it follows that corresponding coefficients and disturbances are equal. This implies immediately that $u_2' = v_2$ and that $b_3 = r$. Moreover, $a_2' + ra_1 = a_1$, making $a_2' = (1 - r)a_1$. Similarly one can show that $b_2' = (1 - r)b_1$. Substituting these values into Equation (2′) gives

$$\begin{aligned} \text{Success} = (1 - r)a_1 + (1 - r)b_1 \text{ (Grades)} \\ + r \text{ (Admission Rating)} + v_2 \end{aligned} \quad (4')$$

Thus, in this example, the coefficient for Grades falls between its coefficients in equations (4) and (4′)—that is, between b_1 and $(1 - r)$ b_1. As selection becomes more stringent, the latter term is approached as a limit; and as the committee makes better and better use of the unmeasured variables, r tends to 1, making the estimated coefficient zero. Of course, the higher the r, the more effective the committee, in the sense that it makes use of factors which could not be used in a mechanical selection procedure, since they are not quantitatively measured. The striking consequence of this analysis is that if a committee weights measured factors perfectly, the better the job they do assessing *unmeasured* factors, the worse they will look. The more unmeasured variables they successfully employ, the lower the coefficient of Grades in the admitted group. That is, increases in r reduce the Grades coefficient.

In fact, suppose that the committee successfully employs all the unmeasured factors, so that Admission Rating is a perfect predictor of subsequent success. This means that Admission Rating = Success and $r = 1$. Then as selection becomes more stringent, the Grades coefficient in the admitted group will tend to zero. The factor used by the committee in making its perfect decisions will appear to have no effect at all on actual success in the program. Since its coefficient is zero, its explained variance will be zero and its correlation with success will be zero. A perfectly efficient committee will thus be evaluated as completely incompetent.

In summary, *low coefficients or correlations of measured factors with outcomes in censored samples give no evidence that a selection procedure performs badly*. When selection is stringent and the selection mechanism makes good use of unmeasured factors, the resulting coefficients and correlations will appear low regardless of their true impact. Correlations are likely to be particularly low, since they are always reduced in subsets with restricted variance on the independent variables. Censoring reduces them still further. Thus low correlations of grades or GRE scores with success in highly selected populations give no evidence that grades and GRE scores do not predict success. Low correlations of prior conviction record or nature of the charge with bail jumping or rearrest give no evidence that the first two factors do not predict behavior or that judges are wrong to use them in their decisions. Censored samples studied with ordinary statistical techniques permit no such inferences.

CONVENTIONAL TECHNIQUES FOR ESTIMATING AND REDUCING BIAS IN CENSORED SAMPLES

Survey researchers frequently must make do with heavily censored samples comprising just a fraction of the intended respondents. To judge the effect of biased sampling, they traditionally compare the measured characteristics of the sample to those of the target population. If females, Republicans, blacks, and the wealthy are all represented in the sample in approximately their population proportions, the low response rate is said not to matter, since a "representative" sample has been obtained. On the other hand, if certain demographic categories are underrepresented in the sample, their numbers may be weighted statistically to compensate for the shortfall. In effect, weighting is used as a correction for the biases in a censored sample.

Consider first the case in which no weighting appears to be necessary because the sample is demographically satisfactory. Unfortunately, nothing in the demographics guarantees consistent censoring. The unmeasured factors that influence selection to the sample may still be correlated with the unmeasured variables that influence a particular dependent variable. For example, if a researcher is interested in explaining political participation (voting, working in a campaign, joining a demonstration), the citizen's interest in politics is likely to be a prominent cause of behavior. Political interest can never be measured precisely, however, and it is likely to influence both participation in politics and participation in the survey. A sample may contain a proportionate number of blacks, Jews, and Republicans; but if these people are disproportionately interested in political life, ordinary statistical procedures will fail to give results applicable to the general population. Not only will political participation rates be exaggerated (an obvious result and one that the researcher is likely to notice), but, more important, the interrelationships among the variables will be altered. Positive relationships may appear negative and vice versa. By itself, a sample's representativeness on a few measured variables gives little or no evidence that subsequent statistical computations can be believed. Faith is warranted only when a researcher knows that unmeasured factors influencing selection to the sample do not also influence the behavior being explained.

When the demographics of a sample do not match those of the population, weighting the sample will restore the apparent representativeness of the data. In most cases, perhaps, the resulting estimates will be better than their unweighted counterparts, but even this much is not certain. Suppose a researcher wishes to estimate the political participation rates of a sample of black and white Americans. The whites respond in large numbers to the survey, but the blacks largely escape the sample. The few blacks who do respond participate actively in politics, let us say, while those who refuse resemble the whites in their political activity.

In this study, the unweighted estimate of the participation rate in the population would ignore the missing blacks, in effect replacing them with whites. For an estimate of participation, this replacement causes no difficulties, since the whites are similar to the missing blacks. By contrast, the weighted estimate essentially would count the highly participant blacks several times to replace the less active members of their race who did not enter the sample. The result would be an overestimate of participation. Yet the researcher is likely to look favorably on this substitution, since she knows nothing of the missing blacks. What she does see is the large differences in participation *in the sample* between the highly participant blacks and the less participant whites. Adjusting for race will seem attractive. The effect of weighting, however, is to worsen the estimate of the population's participation in politics. Thus reweighting a sample need not improve statistical procedures; it may make them worse instead. As before, the culprit is the set of unmeasured factors influencing both political participation and selection to the sample. If the same factors that get blacks into the survey also encourage political activity, then weighting may be worse than doing nothing.

In opinion surveys or in any of the other censored-sample situations discussed in this chapter, then, the conditions for eliminating inconsistency are frequently unmet. The missing cases typically cause biases—biases that neither statistical controls nor sample weighting can eliminate. What can be done when a researcher knows that a sample has been censored in a fashion likely to induce inferential errors? Traditional statistical procedures fail because they assume that conditional on the independent variables, the sample has been chosen randomly. Ignoring the nature of the sampling creates the distortions. It follows that explicit modeling of the data selection process might provide the information needed to correct the errors. The next chapter shows how this can be done.

Appendix

This appendix derives the asymptotic biases (inconsistencies) in regression coefficients estimated on a censored sample when all variables are normally distributed. The first lemma proves some standard results for censored multivariate normal distributions that will be needed below.

Lemma 1. Let u_1, u_2, and u_3 be normally distributed random variables with mean zero and respective variances $\sigma_1^2 = 1$, σ_2^2, and σ_3^2 and covariances σ_{jk}. Let $\varphi(\cdot)$ and $\Phi(\cdot)$ be the density and distribution function of a standard normal variable. Then:

(a) $E(u_1|u_1 \geq z) = \lambda > z$, where $\lambda = \varphi(z)/[1 - \Phi(z)]$.

(b) $\text{var}(u_1|u_1 \geq z) = h$, where $h = 1 + z\lambda - \lambda^2$.

(c) $E(u_2|u_1 \geq z) = \rho_{12}\sigma_2\lambda$, where the correlation $\rho_{ij} = \sigma_{ij}/\sigma_i\sigma_j$.

(d) $\text{var}(u_2|u_1 \geq z) = \sigma_2^2[1 - (1 - h)\rho_{12}^2]$.

(e) If $\sigma_{jk} \neq 0$, then
$\text{cov}(u_j, u_k|u_1 \geq z) = h\sigma_{jk} + (1 - h)\sigma_{jk}(\rho_{jk} - \rho_{1j}\rho_{1k})/\rho_{jk}$.
(If $\sigma_{jk} = 0$, the same expression holds if σ_{jk}/ρ_{jk} is interpreted as $\sigma_j\sigma_k$.)

(f) $0 < h < 1$ for finite z.

Proof:

(a)
$$E(u_1|u_1 \geq z) = \frac{\int_z^\infty x\varphi(x)dx}{\text{Prob}\,(x \geq z)} = \frac{\int_z^\infty xe^{-x^2/2}\,dx}{\sqrt{2\pi}[1 - \Phi(z)]} \tag{A-1}$$

Make the change of variable $y = x^2/2$, which is absolutely continuous. Then (A-1) becomes:

$$\frac{1}{\sqrt{2\pi}[1 - \Phi(z)]} \int_{z^2/2}^{\infty} e^{-y}\, dy = \frac{1}{\sqrt{2\pi}[1 - \Phi(z)]} e^{-z^2/2}$$
$$= \frac{\varphi(z)}{1 - \Phi(z)}$$

Clearly $\lambda > z$ by the definition of λ.

(b) $$E(u_1^2 | u_1 \geq z) = \frac{1}{\sqrt{2\pi}[1 - \Phi(z)]} \int_z^{\infty} x^2 e^{-x^2/2}\, dx \qquad \text{(A-2)}$$

The integrand may be written as

$$\frac{-x \partial e^{-x^2/2}}{\partial x}$$

and so integration by parts gives the following for (A-2):

$$\frac{1}{\sqrt{2\pi}[1 - \Phi(z)]} [-xe^{-x^2/2} + \textstyle\int e^{-x^2/2}\, dx]\Big|_z^{\infty}$$
$$= \frac{z\varphi(z) + 1 - \Phi(z)}{1 - \Phi(z)} \qquad \text{(A-3)}$$
$$= z\lambda + 1$$

using $\lim_{z \to \infty} z\varphi(z) = 0$.[2]

Finally from (A-1) and (A-3):

$$\text{var}(u_1 | u_1 \geq z) = E(u_1^2 | u_1 \geq z) - E^2(u_1 | u_1 \geq z)$$
$$= 1 + z\lambda - \lambda^2$$

(c) Since for normally distributed variables,

$$E(u_2 | u_1 = z) = b_{21} E(u_1 | u_1 = z)$$

with $b_{21} = \sigma_{12}/\sigma_1^2 = \rho_{12}\sigma_2/\sigma_1$, which is not a function of z, the result follows from part (a) and $\sigma_1 = 1$.

---

[2]For $z > 1$, $z\varphi > 0$ and $\partial z\varphi/\partial z = \varphi\ (1 - z^2) < 0$. Thus $z\varphi$ is bounded below and monotonically decreasing; hence it has a limit as $z \to \infty$. But $z\varphi$ is the integrand in the expression for the mean of a normal distribution, and since that integrand has a limit, the limit must be zero if the mean is to exist.

(d) The regression of u_2 on u_1 is $u_2 = \rho_{12}\sigma_2 u_1 + v_{21}$, where v_{21} is independent of u_1 and $\mathrm{var}(v_{21}) = \mathrm{var}(v_{21}|u_1 = z) = (1 - \rho_{12}^2)\sigma_2^2$. Since the variance of v_{21} does not depend on u_1, then, using part (b):

$$\begin{aligned}\mathrm{var}(u_2|u_1 \geq z) &= \rho_{12}^2\sigma_2^2 \,\mathrm{var}(u_1|u_1 \geq z) + (1 - \rho_{12}^2)\sigma_2^2 \\ &= \sigma_2^2[1 - (1 - h)\rho_{12}^2]\end{aligned}$$

(e) Write $u_j = \rho_{1j}\sigma_j u_1 + v_{j1}$, so that $\mathrm{cov}(v_{j1}, v_{k1}|u_1 = z) = \mathrm{cov}(v_{j1}, v_{k1}) = \sigma_{jk} - \rho_{1j}\rho_{1k}\sigma_j\sigma_k = \sigma_{jk}(\rho_{jk} - \rho_{1j}\rho_{1k})/\rho_{jk}$. Then

$$\begin{aligned}\mathrm{cov}(u_j, u_k|u_1 \geq z) &= \rho_{1j}\rho_{1k}\sigma_j\sigma_k \,\mathrm{var}(u_1|u_1 \geq z) \\ &\quad + \mathrm{cov}(v_{j1}, v_{k1}) \\ &= \sigma_{jk}\rho_{1j}\rho_{1k}h/\rho_{jk} + \sigma_{jk}(\rho_{jk} - \rho_{1j}\rho_{1k})/\rho_{jk} \\ &= h\sigma_{jk} + (1 - h)\sigma_{jk}(\rho_{jk} - \rho_{1j}\rho_{1k})/\rho_{jk}\end{aligned}$$

(f) First, since the variance in part (b) cannot vanish, $h > 0$. Next, since $\lambda > 0$ and $\lambda > z$,

$$h = 1 + z\lambda - \lambda^2 < 1 + \lambda^2 - \lambda^2 = 1$$

The next lemma shows that the variance of a censored normal distribution declines monotonically to zero as censoring becomes more severe.

Lemma 2. (a) $\lim_{z\to\infty} h = \lim \mathrm{var}(u_1|u_1 \geq z) = 0$; (b) $\partial h/\partial z < 0$.

Proof:

(a) Write h as:

$$h = \frac{(1 - \Phi)^2 + z\varphi(1 - \Phi) - \varphi^2}{(1 - \Phi)^2}$$

Using L'Hôpital's rule twice with respect to Φ,[3]

$$\lim_{z\to\infty} h = \lim_{\Phi\to 1} h = 1 - \lim_{\Phi\to 1} \frac{z(1 - \Phi)}{\varphi} \qquad \text{(A-4)}$$

---

[3]The following results are needed: as $z \to \infty$, $\lim z\varphi = 0$ and $\lim z(1 - \Phi) = \lim z^2(1 - \Phi) = 0$. The first is proved in note 2. Now from Lemma 1 (a), $\varphi > z(1 - \Phi)$ and hence $z\varphi > z(1 - \Phi)$ and $z\varphi > z^2(1 - \Phi)$ for $z > 1$. Since neither of the latter right-hand terms is ever negative for $z > 0$, it follows that their limit must be zero because $z\varphi$ goes to zero.

But a final application of L'Hôpital's rule to the last limit on the right in (A-4) gives:

$$\lim \frac{z(1-\Phi)}{\varphi} = 1 - \lim\left(\frac{1}{z\lambda}\right) = 1$$

since $\lambda > z$ and $z \to \infty$ as $\Phi \to 1$. Hence from (A-4), $\lim h = 0$.

(b) It is straightforward to show that

$$\frac{\partial h}{\partial \Phi} = \frac{h - (\lambda - z)^2}{1 - \Phi} \tag{A-5}$$

and

$$\frac{\partial^2 h}{\partial \Phi^2} = \frac{2(\partial h / \partial \Phi)}{(1 - \Phi)} + p \tag{A-6}$$

where $p = 2(\lambda - z)h/[\varphi(1 - \Phi)] > 0$, since $\lambda - z$, h, φ, $1 - \Phi > 0$.

Inspection of the continuous second derivative (A-6) shows that if $\partial h / \partial \Phi$ becomes nonnegative at some point Φ_0, equation (A-6) will be positive so that $\partial h / \partial \Phi$ will be increasing at Φ_0. But then for all $\Phi > \Phi_0$, we have $\partial^2 h / \partial \Phi^2 > 0$ and $\partial h / \partial \Phi > 0$. Hence h increases monotonically beyond Φ_0 and since $h > 0$, we have $\lim h > 0$, contrary to part (a). Therefore $\partial h / \partial z = \varphi \partial h / \partial \Phi < 0$.

Now consider a two-equation system in which the first dependent variable determines whether the second will be observed. The first equation is of the probit form:

$$y_1^* = X_1 \beta_1 + u_1 \tag{A-7}$$

where y_1^* is an unobserved n- dimensional vector of "normits," X_1 is an $n \times k_1$ matrix of observations on k_1 exogenous (independent) variables with mean zero, β_1 is an unknown vector of k_1 coefficients, and u_1 is the disturbance term. Corresponding to each element y_{1i}^* of y_1^* is an observed variable y_{1i}:

$$y_{1i} = \begin{cases} 1 & \text{if } y_{1i}^* \geq d \\ 0 & \text{otherwise} \end{cases}$$

where d is a constant to be determined. This formulation differs from the usual probit structure only in standardizing the independent variables to mean zero and suppressing the intercept.

It is also assumed that:

(i) The elements of u_1 are distributed independently and normally with mean zero and variance $\sigma_1^2 = 1$.

(ii) The rows X_{1i} of X_1 are each independent realizations of a multivariate normal distribution with mean zero and constant positive definite covariance matrix G. The covariance of X_1 with u_1 is zero.

These two assumptions imply that:

(iii) $\text{plim } X_1'u_1/n = 0$.

(iv) Asymptotically X_1 is of full rank k_1 in probability.

(v) Asymptotically the coefficient vector β_1 is identified in probability. (The conditions for probit coefficients to be identified appear in Haberman, 1974, pp. 314–21.)

(vi) Let X_{1i} be the ith row of X_1, $\Phi_i = \Phi(X_{1i}\beta_1)$, $\varphi_i = \varphi(X_{1i}\beta_1)$, $\lambda_i = \varphi_i/(1 - \Phi_i)$, and $h_i = 1 + X_{1i}\beta_1\lambda_i - \lambda_i^2$. By Lemma 1(f), $0 \leq h_i\,(1 - \Phi_i) < 1$; hence the unconditional expectation of $h_i(1 - \Phi_i)$ exists.

The second equation, the structure of primary interest, is:

$$y_2 = X_2\beta_2 + u_2 \tag{A-8}$$

where y_2 is an n-dimensional vector of observations on the dependent variable, X_2 is an $n \times k_2$ matrix of observations on k_2 independent variables standardized to mean zero, β_2 is a coefficient vector to be estimated, and u_2 is the disturbance vector.

Equation (A-8) is observed only when $y_{1i} = 1$. By convention, it is the first m rows of (A-8) that are observed:

$$y_{21} = X_{21}\beta_2 + u_{21} \tag{A-9}$$

It is assumed that:

(vii) The elements of u_2 are distributed independently and normally with mean zero and unknown variance σ_2^2. Corresponding elements of u_1 and u_2 have covariance σ_{12}.

(viii) The rows of X_2 are independent realizations of a multivariate normal distribution with mean zero and constant positive definite variance-covariance matrix J. The covariances of X_2 with u_1 and u_2 are zero.

These two assumptions imply that:

(ix) $\operatorname{plim} X_2'u_2/n = 0$, $\operatorname{plim} X_2'u_1/n = 0$.

Now suppose that OLS is applied to equation (A-9), that is, to those m observations for which data on y_2 are available. The resulting estimate of β_2 is:

$$\hat{\beta}_2 = (X_{21}'X_{21})^{-1}X_{21}y_{21} \qquad \text{(A-10)}$$

To find the probability limit of this estimate, note first that each element of $\operatorname{plim} X_{21}'X_{21}/n$ is a mean of covariances, the ith of which is observed conditional on $y_{1i}^* \geq d$. Now under the preceding assumptions, conditional on the ith data point being observed, the covariance between the jth and kth independent variables for that observation may be written, using Lemma 1(e), as:

$$h_i\sigma_{jk}^* + \frac{(1 - h_i)\sigma_{jk}^*(\rho_{jk} - \rho_{1j}\rho_{1k})}{\rho_{jk}}$$

where σ_{ij}^* is the unconditional covariance, ρ_{1j} and ρ_{1k} are the zero-order correlations of the two independent variables with y_1^*, and ρ_{jk} is their zero-order correlation with each other. If the cross-product of two independent variables is set to zero when the data point is not observed, the unconditional covariance in a censored sample is simply the conditional covariance multiplied by $(1 - \Phi_i)$, which is the probability of being observed.

Now clearly $\operatorname{plim} X_2'X_2/n$ is the matrix whose typical element is σ_{ij}^*. Next, it is easy to show that:

$$\operatorname{plim} \frac{[X_2'X_2 - X_2'y_1^*(y_1^{*\prime}y_1^*)^{-1}y_1^{*\prime}X_2]}{n}$$

is the matrix whose typical element is

$$\frac{\sigma_{jk}^*(\rho_{jk} - \rho_{1j}\rho_{1k})}{\rho_{jk}}$$

Using these results in the conditional covariance matrix just given and setting

$$\mu_1 = \operatorname{plim} \frac{\sum_{i=1}^{n} h_i(1 - \Phi_i)}{n};$$

$$\mu_2 = \operatorname{plim} \frac{\sum_{i=1}^{n} (1 - h_i)(1 - \Phi_i)}{n}$$

yields:

$$\operatorname{plim} \frac{X_{21}'X_{21}}{n} = \mu_1 \operatorname{plim} \frac{X_2'X_2}{n} + \mu_2 \operatorname{plim} \frac{X_2'MX_2}{n} \tag{A-11}$$

where $M = I - y_1^*(y_1^{*\prime}y_1^*)^{-1}y_1^{*\prime}$. All these probability limits exist because their arguments are means of independent and identically distributed variables with finite expectations. (See Assumption (vi).)

Similarly:

$$\operatorname{plim} \frac{X_{21}'y_{21}}{n} = \mu_1 \operatorname{plim} \frac{X_2'y_2}{n} + \mu_2 \operatorname{plim} \frac{X_2'My_2}{n} \tag{A-12}$$

Thus, setting $\delta_1 = \mu_1/(\mu_1 + \mu_2)$, equations (A-11) and (A-12) imply:

$$\operatorname{plim} \hat{\beta}_2 = \operatorname{plim}[\delta_1 X_2'X_2 + (1 - \delta_1)X_2'MX_2]^{-1} \times [\delta_1 X_2'y_2 + (1 - \delta_1)X_2'My_2] \tag{A-13}$$

Since $\mu_1, \mu_2 > 0$, then $0 \leq \delta_1 \leq 1$, so that $\hat{\beta}$ is a matrix weighted average of $\operatorname{plim}(X_2'X_2)^{-1}X_2'y_2 = \beta_2$, the true value, and $\bar{\beta}_2 = \operatorname{plim}(X_2'MX_2)^{-1} X_2'My_2$, the coefficient vector of X_2 when y_2 is regressed on X_2 and y_1^*. (This interpretation of $\bar{\beta}_2$ follows easily from partitioning the matrix of independent variables into X_2 and y_1^* in the latter regression.)

The ith data point is observed if $y_{1i}^* \geq d$ or, equivalently, if the standard normal variable $y_{1i}^* - X_{1i}\beta_1 \geq d - X_{1i}\beta_1$. As $d \to \infty$, fewer data points will be selected. The following lemma shows that as selection becomes more stringent in this fashion, the weight δ_1 approaches the limit zero.

Lemma 3. Assume the model of (A-7) and (A-8) under assumptions (i), (ii), (vii), (viii). Define $\hat{\beta}$ as in (A-10) and $\bar{\beta}$ as below (A-13). Then

$$\lim_{d\to\infty} \operatorname{plim} \hat{\beta} = \bar{\beta}$$

Proof: By (A-13), we need only show

$$\lim_{d\to\infty} \delta_1 = \lim_{d\to\infty} \frac{\operatorname{plim} \sum_{i=1}^{n} h_i \frac{(1 - \Phi_i)}{n}}{\operatorname{plim} \sum_{i=1}^{n} \frac{(1 - \Phi_i)}{n}} = 0$$

But since h_i and $(1 - \Phi_i)$ exceed zero, the ratio is bounded above by the largest h_i, which goes to zero by Lemma 2.

The main result can now be demonstrated.

Theorem. In the model given by (A-7) and (A-8), assume (i), (ii), (vii), and (viii), and define the OLS estimate $\hat{\beta}_2$ as in (A-10) and $\bar{\beta}_2$ as below (A-13). Then there exists a nonsingular matrix A such that if X_2 is transformed to AX_2, each element of the resulting coefficient vector, plim $A^{-1}\hat{\beta}_2$, is a simple linear combination of the corresponding elements of $A^{-1}\beta_2$ and $A^{-1}\bar{\beta}_2$; moreover, as selection becomes more stringent ($d \to \infty$), plim $A^{-1}\hat{\beta}_2$ approaches $A^{-1}\bar{\beta}$ in each coefficient.

Proof:

Since $X_2'X_2/n$ and $[X_2'X_2 - X_2'y_1^*(y_1^{*\prime}y_1^*)^{-1}\, y_1^{*\prime}X_2]/n$ are positive definite (the latter because y_1^* and X_2 cannot be collinear by assumption viii), there exists a nonsingular matrix A such that

$$\frac{A'X_2'X_2A}{n} = I \tag{A-14}$$

and

$$\frac{A'[X_2'X_2 - X_2'y_1^*(y_1^{*\prime}y_1^*)^{-1}\, y_1^{*\prime}X_2]\, A}{n} = \Lambda \tag{A-15}$$

a diagonal matrix with positive diagonal elements λ_j (Dhrymes, 1970, pp. 481–82).

Now (A-14) and (A-15) with (A-13) imply

$$\text{plim}\, A^{-1}\hat{\beta}_2 = \text{plim}[\delta_1 I + (1 - \delta_1)\Lambda]^{-1} \times \frac{[\delta_1 A'X_2'y_2 + (1 - \delta_1)A'X_2'My_2]}{n}$$

so that when $\delta_1 = 1$,

$$\text{plim}\, A^{-1}\hat{\beta}_2 = \text{plim}\, \frac{A'X_2'y_2}{n} = \text{plim}\, \frac{A'X_2'(X_2AA^{-1}\beta_2 + u)}{n} = A^{-1}\beta_2$$

and as $\delta_1 \to 0$,

$$\text{plim}\, A^{-1}\hat{\beta}_2 \to \text{plim}\, \frac{\Lambda^{-1}A'X_2'My_2}{n} = A^{-1}\bar{\beta}_2$$

Hence:

$$\text{plim}\, A^{-1}\hat{\beta}_2 = [\delta_1 I + (1 - \delta_1)\Lambda]^{-1} \times [\delta_1 A^{-1}\beta_2 + (1 - \delta_1)\Lambda\, A^{-1}\bar{\beta}_2]$$

Since the inverse is diagonal, any element of plim $A^{-1}\hat{\beta}_2$, say the first, can be written in an obvious notation as:

$$\text{plim}(A^{-1}\hat{\beta}_2)_1 = \frac{\delta_1(A^{-1}\beta_2)_1 + (1 - \delta_1)\lambda_1(A^{-1}\bar{\beta}_2)_1}{\delta_1 + (1 - \delta_1)\lambda_1} \qquad \text{(A-16)}$$

which is clearly a linear combination of the first elements of $A^{-1}\beta_2$ and $A^{-1}\bar{\beta}_2$. Moreover, by Lemma 3 the weight on $A^{-1}\bar{\beta}_2$ increases to 1 as d approaches infinity.

In the bivariate regression case (with variables standardized to mean zero and the intercept omitted), the same linear combination and limit properties hold for the single coefficient. Moreover, since A^{-1} is a scalar in that instance and can be factored out of both sides of equation (A-16), both properties hold for the original untransformed coefficient as well.

The corollary below gives the OLS bias for the Tobit setup, in which y_{2i}, the dependent variable in the second structural equation, is observed only when it exceeds a constant d. That is, Tobit is the special case of censored sampling for which $y_1^* = y_2$, so that (A-7) and (A-8) are identical. The corollary shows that in this instance, OLS attenuates all coefficients by the same fraction, a result first obtained by Goldberger (1978). It is also demonstrated that more stringent selection drives all coefficients towards the limit zero. Hence the R^2 goes to zero as well.

Corollary (Tobit Case). In the model given by (A-7) and (A-8), assume $y_1^* = y_2$, and define the OLS estimate $\hat{\beta}_2$ as in (A-10). Then under assumptions (i), (ii), (vii), (viii), plim $\hat{\beta}_2 = k\beta_2$, with $0 < k < 1$. Moreover, as selection becomes more stringent ($d \to \infty$), plim $\hat{\beta}_2 \to 0$.

Proof: Applying (A-13) to this case yields:

$$\text{plim}\, \hat{\beta}_2 = \text{plim}[\delta_1 X_2' X_2 + (1 - \delta_1) X_2' M X_2]^{-1} \times [\delta_1 X_2 y_2 + (1 - \delta_1) X_2' M y_2], \qquad \text{(A-17)}$$

where here $M = I - y_2(y_2' y_2)^{-1} y_2'$

Noting that $X_2' M y_2 = 0$ and rearranging:

$$\begin{aligned}\text{plim } \hat{\beta}_2 &= \text{plim } \delta_1[X_2'X_2 - (1-\delta_1)X_2'y_2(y_2'y_2)^{-1}y_2'X_2]^{-1}X_2'y_2 \\ &= \text{plim } \delta_1(X_2'X_2)^{-1}[(X_2'X_2)^{-1} - (1-\delta_1)(X_2'X_2)^{-1}X_2'y_2 \\ &\quad \times (y_2'y_2)^{-1}y_2'X_2(X_2'X_2)^{-1}]^{-1}(X_2'X_2)^{-1}X_2'y_2\end{aligned} \tag{A-18}$$

Since $\text{plim } (X_2'X_2)^{-1}X_2'y_2 = \beta_2$:

$$\text{plim } \hat{\beta}_2 = \text{plim } \delta_1(X_2'X_2)^{-1}[(X_2'X_2)^{-1} - (1-\delta_1)\beta_2(y_2'y_2)^{-1}\beta_2']^{-1}\beta_2 \tag{A-19}$$

Now for invertible matrix A, constant α, and vector β (Dhrymes, 1978, p.459):

$$(A + \alpha\beta\beta')^{-1} = A^{-1} - cA^{-1}\beta\beta'A^{-1}, \tag{A-20}$$

with $c = \alpha/(1 + \alpha\beta'A^{-1}\beta)$. Applying this to the expression in brackets in (A-19) and setting $\rho^2 = \beta_2'X_2'X_2\beta_2/y_2'y_2$ (the squared correlation over the full population for the outcome equation) yields:

$$\begin{aligned}\text{plim } \hat{\beta}_2 &= \delta_1\left[I + \frac{(1-\delta_1)\rho^2}{1-(1-\delta_1)\rho^2} I\right]\beta_2 \\ &= \frac{\delta_1}{1-(1-\delta_1)\rho^2}\beta_2,\end{aligned} \tag{A-21}$$

which, apart from notational differences, is Goldberger's result. Since both ρ^2 and δ_1 are strictly positive but less than unity, it is straightforward to show that (A-21) is of the form $k\beta_2$ with $0 < k < 1$.

The second part of the Corollary follows from the Theorem and the fact that here, $\bar{\beta}_2 = \text{plim } (X_2'MX_2)^{-1}X_2'My_2 = 0$.

5

Estimating Treatment Effects in Quasi-Experiments: The Case of Censored Data

CORRECTING FOR CENSORING IN THE PROBIT SELECTION CASE

None of the usual statistical techniques, such as cross-tabulation or regression analysis, produces dependable results when a sample has been nonrandomly drawn from its population. As noted in the previous chapter, ignoring the nature of the sample poses formidable dangers to the data analyst: treatments with a positive effect may appear to have weak or even negative impacts, and intuitively plausible adjustments may make matters worse. More powerful statistical methods are required for censored samples.

As in the case of nonrandomized assignment to treatment and control groups, corrections for nonrandom sampling require the modeling of both the process of selection into the sample and the ensuing outcomes. This chapter sets out the mechanics of the appropriate estimators. To keep matters simple, the nonrandomness is initially assumed to be confined to the drawing of the sample. The assignment to treatment and control groups (whether these constitute discrete categories or a continuum) is assumed to be essentially random over the population as a whole. The next chapter discusses briefly how this latter restriction may be relaxed.

Consider first the structure discussed in the last chapter, in which selection into the sample occurs according to the assumptions of the probit model. The selection equation is

$$y^*_{1i} = a_1 + b_{11}x_1 + b_{21}x_2 + \cdots + b_{k1}x_k + u^*_{1i} \tag{1}$$

and the ith data point enters the sample of outcomes if $y_{1i}^* \geq 0$. Here the x_j's are independent (exogenous) variables, the b_{j1}'s are unknown coefficients, a_1 is an intercept, and u_{1i}^* is a disturbance term assumed to be a standard normal random variable distributed independently of the x_j's and independently across the observations. The normit y_{1i}^* is a hypothetical variable and is not observed. The variable y_{1i} is observed, however. It equals 1 if y_{1i}^* exceeds zero (in which case the outcome equation is observed) and equals zero otherwise (unobserved outcome equation).

This specification says simply that certain observed factors linearly and additively influence selection to the sample. In the example of the previous chapter, "selection" might mean admission to graduate school. In that case the independent variables would represent college grades, strength of letters of recommendation, scores on standardized tests, and so on, each of which is measured for *every* applicant. If no such variables are measured, the analysis ordinarily cannot proceed. If nothing is known about the selection process, or if no data are available from the group not selected, no adjustment for censoring is possible. Not *all* factors influencing the admission decision need be measurable, however; as usual, unmeasured factors uncorrelated with the included independent variables may be consigned to the disturbance.

The second equation, which is of principal interest, explains outcomes as a function of one or more treatments and perhaps additional variables regarded as control factors. This specification applies in principle to all observations but is observed only for those in the censored sample. The dependent variable y_{2i} is the outcome for the ith observation, the x_j's are independent variables, the b_{j2}'s are unknown coefficients, and u_{2i} is a disturbance term. In the full, uncensored sample, u_{2i} is assumed to be normally distributed with mean zero and constant variance, distributed independently of the x_j's and other disturbances. Then the outcome equation is

$$y_{2i} = a_2 + b_{12}x_{1i} + b_{22}x_{2i} + \cdots + b_{K2}x_{Ki} + u_{2i} \qquad (2)$$

In the admissions example, the independent variables in this equation might be grades, test scores, or other proxies for ability and motivation; the dependent variable would be a measure of success in the graduate program. Data on success would be available only for matriculants, but the equation would have predicted the success of the other applicants just as well if they had been admitted. For a censored sample to be statistically manageable, it is essential that at least one independent variable appearing in the selection equation (1) not be entered in the outcome equation (2). That is, the outcome specification must embody an exclusion restriction; the researcher must know that some factor influencing selection makes no difference in outcomes. For example, the admissions committee might know that alumni children get preference in admissions but perform neither better nor worse than other students with the same qualifications. Although this restriction is not strictly required for identification, without it the estimation procedure set out below will work poorly in practice due to near collinearity.

The need for a special estimator arises from the nature of the two disturbances. In most quasi-experiments, u^*_{1i} and u_{2i} are correlated. In that case, the appendix to the previous chapter showed that in the censored sample the disturbance term u_{2i} has neither mean zero nor zero correlation with the independent variables, even though it has both properties in the full sample. Ordinary least squares applied to the outcome equation (2) gives erroneous results for just this reason.

To correct the misspecification, the expected value of u_{2i} under censoring, denoted λ, is estimated and added to the regression equation as an additional variable. Asymptotically, this procedure removes the correlated part of u_{2i} from the disturbance, creating a new disturbance term that is uncorrelated with the independent variables. (See the appendix to the previous chapter.) Consistent estimates now result from applying least squares to the new equation.

The computational procedure is as follows (Heckman, 1976, 1979):

A. Apply probit analysis to the selection equation (1), using the full sample of observations. Use the resulting probit coefficients and the independent variables to forecast y^*_{1i} in the usual way and denote each forecast by $\hat{y}^*_{1i}$. Thus if estimated probit coefficients are denoted by $\hat{a}_1$ and $\hat{b}_{j_i}$

$$\hat{y}^*_{1i} = \hat{a}_1 + \hat{b}_{11}x_1 + \hat{b}_{21}x_2 + \cdots + \hat{b}_{k1}x_k$$

Let the density of a standard normal curve at the point $\hat{y}^*_{1i}$ be φ_i, and let the cumulative distribution function of the same curve at the same point be Φ_i. Then $\lambda = \varphi_i/(1 - \Phi_{1i})$.

B. Add λ to the outcome equation (2) as an additional variable. Apply ordinary least squares, using just the censored sample, to the resulting equation:

$$y_{2i} = a_2 + b_{12}x_{1i} + b_{22}x_{2i} + \cdots + b_{K2}x_{Ki} + b_{K+1,2}\lambda + v_{2i}$$

The result is consistent estimates of the coefficients in the outcome equation.

C. The standard errors of the coefficients produced by the least squares calculations in step B are incorrect in general. Formulas for the true sampling errors may be found in Heckman (1979). (If the censored sample is small relative to the full sample and/or the coefficient on λ is no more than a small fraction of the standard deviation of the residuals, the estimated standard errors produced by the OLS calculations will generally provide only a modest proportional underestimate of the true values.)

When the outcome measure is continuous, Heckman's method corrects for censoring bias so long as disturbances are normally distributed. In some applications, however, outcomes may be discrete. As noted earlier, the subjects of a study frequently are classed as "successes" or "failures," so that experimental results are dichotomous. One might hope that Heckman's method would extend to this case—that is, λ would be estimated as before and it would be added as an additional variable to a probit equation that modeled the dichotomous experimental outcomes. This approach fails, however. It can be shown that the resulting disturbances in the second probit equation are no longer normally distributed, so that probit analysis is inconsistent. Thus Heckman's method is restricted to the case of continuous outcomes.

CORRECTING FOR CENSORING IN THE LINEAR SELECTION CASE

The computational inconvenience of selection models based on probit theory motivates another correction technique. In this approach, selection to the sample is based on the linear probability model (introduced in Chapter 3). That is, the dichotomous dependent variable in the selection equation, y_{1i}, is assumed to be a linear function of certain exogenous variables:

$$y_{1i} = a_1 + b_{11}x_1 + b_{21}x_2 + \cdots + b_{k1}x_k + u_{1i} \quad (3)$$

Ordinary least squares applied to this equation is unbiased and consistent, and generalized least squares ("Goldbergerizing," as explained in Chapter 3) is consistent and asymptotically efficient. Then the ith observation on the outcome equation enters the sample if $y_{1i} = 1$ and does not enter if $y_{1i} = 0$.

The outcome of the quasi-experiment is also assumed to depend linearly on exogenous factors, just as in the probit case:

$$y_{2i} = a_2 + b_{12}x_{1i} + b_{22}x_{2i} + \cdots + b_{K2}x_{Ki} + u_{2i} \quad (2)$$

If outcomes vary continuously, it may be reasonable to assume that in the full sample the disturbance term, u_{2i}, is distributed with mean zero and fixed variance. Once again, it is to be expected that some unobserved quantities will influence both selection and outcomes, so that u_{1i} and u_{2i} will be correlated. One must also assume that the outcome equation contains an exclusion restriction—an assumption now required if the estimation procedure is to work at all.

To complete the assumptions, the form of the covariance between the disturbances must be specified. In censored-sample theory, this structural feature critically affects the statistical properties of the model, and its specification deserves special attention. In nearly all econometric theory, when two disturbance terms are normally distributed, their variances and the covariances between them are assumed constant across observations. Heckman takes this approach in his analysis of the probit selection model, for example. This postulate implies that both the correlation between the disturbances and their regression function are also fixed. The convenient properties of bivariate normal distributions under these conditions then combine to make Heckman's estimator manageable.

When the selection equation is a linear probability model, a variety of assumptions can be made about the structure of the disturbances in the model. Among these, it is convenient to choose one that parallels the probit selection model in its underlying theoretical ideas. In that manner, one derives an estimator simpler than the probit method but close to it in assumptions and behavior. Just as one often uses a linear probability model rather than the probit model in estimating an ordinary single-equation model—essentially because the linear probability model is simpler and cheaper to compute and rarely misleads—so also a linear probability approach to censored samples confers some of the same benefits.

The linear probability model is usually presented simply as a variant of regression with no attempt to justify it by appeal to an underlying behavioral model. Probit analysis, of course, has such a theoretical rationale, which accounts for some of its appeal. But in fact, one can derive the linear probability model by postulating an underlying regression model whose dependent variable is not observed, just as in the probit case. Suppose that the underlying regression is:

$$y^*_{1i} = a_1 + b_{11}x_1 + b_{11}x_2 + \cdot\cdot\cdot b_{k1}x_k + u_{1i}$$

Here y^*_{1i} is not observed. However, the dichotomous dependent variable y_{1i} is observed and is defined as follows:

$$y_{1i} = 1 \text{ if } y^*_{1i} \geq \tfrac{1}{2}, \text{ otherwise } y_{1i} = 0$$

The specification is then completed by assuming that u^*_{1i} has a uniform distribution over the interval $[-\frac{1}{2}, \frac{1}{2}]$. It is straightforward to show that this setup implies the usual linear probability model specification (3). In other words, the usual linear probability setup holds and the standard techniques apply. The somewhat implausible uniform distribution assumption is justified not by its realism, but by the good approximations to more elaborate estimates that are obtained in practice.

Viewed in this light, the linear probability model and the probit model share nearly the same underlying regression specification. That is, equations (1) and (1′) are identical. The specifications differ only in the distributional assumption on the error term in the underlying regression. The probit model postulates that the underlying disturbance is standard normal; the linear probability model assumes that it is uniform. (The threshold at which y_{1i} switches from 0 to 1 also differs, but that is just a reflection of the difference in scales. Zero on the probit scale and $\frac{1}{2}$ on the linear probability scale both correspond to 50 percent probability.)

With this prologue, a simple correction for selection bias when the sample is censored according to a linear probability model can now be set out. The key assumption is that the regression function relating u_{2i} to u_{1i} is linear with constant coefficient and constant residual variance, just as in the probit selection case. While these assumptions have a certain ad hoc quality when used with a uniform distribution, the result is a very simple selection-bias correction that can be computed using nothing but ordinary regression and that behaves very much like its probit cousin (independently, Olson, 1980; and Achen, 1980). Mathematical details are supplied in the appendix to this chapter.

The calculations for the linear probability correction for selection bias are as follows:

A. In the selection equation (3), set the dependent variable to one if the observation entered the sample and to zero if it did not. Estimate this equation as a linear probability model, using the data from the full sample. (The procedure is given in Chapter 3. Either the one-step or the two-step version of the linear probability model may be used, though the latter is theoretically preferred.)

B. Using the coefficients from the final part of step A, compute forecasts. If any forecasts exceed 0.99 or fall below 0.01, reset them to those limits. Then compute residuals in the usual way by subtracting the forecasts from the original dichotomous dependent variable. Call this residual $\hat{u}_{1i}$. (In all calculations in this step, the original independent and dependent variables are to be used, not the standardized variables used in the second stage of the linear probability computations.)

C. Add $\hat{u}_{1i}$ to the outcome equation as an additional independent variable, obtaining:

$$y_{2i} = a_2 + b_{12}x_{1i} + b_{22}x_{2i} + \cdots + b_{K2}x_{Ki} + b_{K+1,2}\hat{u}_{1i} + v_{2i}$$

Then apply ordinary regression using the censored sample. Consistent estimates of the coefficient result, and their standard errors can be computed by using the formulas in the appendix. As in the probit case, if the coefficient on $\hat{u}_{1i}$ is no more than a small fraction of the residual standard deviation and/or the full sample is large relative to the censored sample, then the standard errors generated by the ordinary regression calculations will usually give only a slight proportional underestimate of the true values.

This procedure differs from Heckman's in two respects. First, this method assumes a linear probability model selection equation whereas Heckman postulates probit selection. In estimating the selection process, then, the two approaches have the usual costs and benefits associated with probit and linear probability models: probit has elegance and somewhat greater plausibility on its side, but the linear probability calculations are simpler, faster, and cheaper, generally with little loss in explanatory power.

Second, and as a consequence of the first difference, the two techniques also differ slightly in their treatment of the outcome equation. Heckman adds λ as an additional variable; the linear proba-

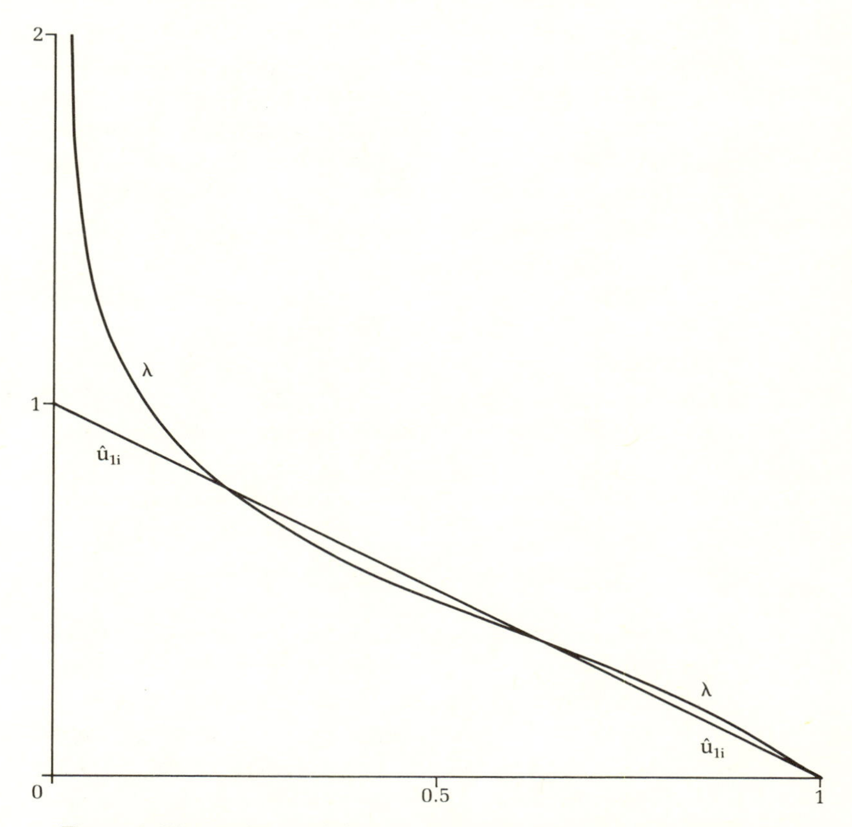

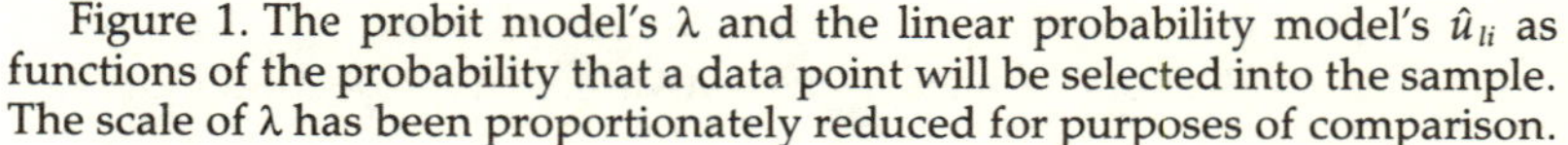

Figure 1. The probit model's λ and the linear probability model's $\hat{u}_{li}$ as functions of the probability that a data point will be selected into the sample. The scale of λ has been proportionately reduced for purposes of comparison.

bility methods adds $\hat{u}_{1i}$. Again the probit calculations sacrifice simplicity to gain credibility, but the linear probability model provides a close approximation to it. Figure 1 shows λ and $\hat{u}_{1i}$ as a function of the probability of selection. The two variables are transformed multiplicatively to the same scale for comparison.[1] (Multiplicative variable transformations leave all other coefficients in a regression unchanged; hence if the two variables are similar on this scale, they will have similar effects in the outcome equation.)

Figure 1 shows that for probabilities of selection exceeding 10 percent, the two variables are quite similar; for probabilities exceeding 20 percent, they are virtually identical.[2] The two specifications differ substantially only for probabilities of selection less than 10 percent, and by definition these are the observations least likely to enter the censored sample. If all selection probabilities are equally likely, for example, those with probabilities below 10 percent will constitute just 1 percent of the censored sample and all those below 20 percent selection probability will number just 4 percent of the sample. Hence in most practical problems there is very little difference on this score between Heckman's method and the simpler linear probability approach.[3]

---

[1]This comparison between u_{1i} and λ assumes that probabilities of selection are the same for each. In practice, the two methods estimate the selection equation differently. Thus selection probabilities are estimated somewhat differently as well, so that the calculations of λ and u_{1i} are based on distinct probabilities. To separate this effect from intrinsic differences between λ and u_{1i}, Figure 2 assumes that selection probabilities are identical for the two variables. Since probit and linear probability forecasts differ little in most applications, this point is a minor one.

[2]If observations of selection probabilities are calculated at intervals of 10 percent between 10 and 100 percent inclusive u_{1i} and λ correlate at 0.989; between 20 and 100 percent they correlate at 0.997.

[3]In the probit selection model, heteroskedasticity occurs after the variable λ has been added to the equation. Specifically, the residual variance increases monotonically as the probability of selection rises. (See the appendix to the previous chapter.) In the linear probability model set out in the appendix to this chapter, the same behavior is exhibited. A plot of the residual variance in the selection equation for both models shows that the linear probability variances are virtually identical to those of the probit model. If the correlation between the two normally distributed disturbances in the probit case is set to 0.5, for example, the linear probability model correlation can be chosen so that its residual variances are within 3 percent either way of the probit model variances for selection probabilities exceeding 50 percent and within 6 percent for other probabilities. Thus in practice the two models are quite similar in this respect also, especially since the observations with higher selection probabilities are more likely to enter the sample.

CORRECTING FOR CENSORING WHEN OUTCOMES ARE DICHOTOMOUS

When outcomes are dichotomous (say, success or failure), the linear probability correction for selection bias outlined above will fail, just as Heckman's estimator does. Regardless of whether the outcome equation is specified as a probit or linear probability model to cope with its dichotomous dependent variable, mechanical application of either Heckman's method or the linear probability technique yields inconsistent coefficient estimates under any plausible assumptions. Dichotomous outcomes require estimation methods that take account of their discontinuous nature.

Modifying Heckman's method to cope with discrete outcomes is impossible if one requires that the estimator be computable with ordinary software of the sort found in major statistical computing packages. The linear probability correction for selection bias can be extended in a plausible way to deal with this case, however. If both the selection and outcome equations are modeled as linear probability equations, the disturbances u^*_{1i} and u^*_{2i} in the two underlying regression models are each uniformly distributed. It is possible to impose a structure on their joint distribution such that the regression function for either one is linear in the other. (See the appendix.) As in any linear probability applications, these distributional assumptions lack a compelling theoretical rationale but imply a convenient estimator which is unlikely to mislead. Here the joint distribution implies that the outcome equation can be estimated consistently in censored samples by adding a single additional term to the equation in the style of the other selection bias estimators outlined above. The equation becomes nonlinear, but it may be estimated relatively simply by nonlinear least squares.

The estimation procedure is as follows:

A. Estimate the selection equation as a linear probability model, using the full sample. Either the one-step or two-step version may be used.

B. Compute the residuals $\hat{u}_{1i}$. This is the same procedure as in the continuous case (step B) given earlier.

C. Let p_{2i} be the right-hand side of the selection equation for the ith observation but without its disturbance term. That is,

$p_{2i} = a_2 + b_{12}x_{1i} + \cdots + b_{K2}x_{Ki}$. Let $q_{2i} = 1 - p_{2i}$ and let v_{2i} be a residual. Then, using just the censored sample, estimate the following equation by nonlinear least squares:

$$y_{2i} = p_{2i} + b_{K+1,2}p_{2i}q_{2i}\hat{u}_{1i} + v_{2i} \quad (4)$$

When this equation is written out in terms of the independent variables and $\hat{u}_{1i}$, it constitutes a nonlinear least squares specification. It may be estimated by using any of the widely available nonlinear regression routines, such as those in the standard statistical computing packages. (Plausible starting values may be obtained by estimating equation (4) using ordinary regression with the product $p_{2i}q_{2i}$ fixed at, say, 0.2.) Nonlinear least squares produces consistent estimates of the coefficients, although the standard errors are erroneous. Formulas for the standard errors, along with a formal statement of the model, are presented in the appendix.[4]

AN EXAMPLE OF SELECTION BIAS CORRECTION

As noted earlier, nonrandomly drawn samples appear routinely in policy analysis. To make clear how selection bias is corrected in practice, an example from the study of pretrial release is presented here. After every arrest, defendants are released on their own recognizance (OR), assigned a bail amount, or, in rare instances, held without bond in "preventive detention" (Bases, 1972). All of the first group and none of the third group go free until trial. Defendants required to post bail, however, may or may not attain release, depending on the size of the bond and their financial resources.

To evaluate a pretrial release system is to ask whether the appropriate defendants are being released (Landes, 1974a; Nagel and Neef,

---

[4]Strictly speaking, the joint distribution of u^*_{1i} and u^*_{2i} is defined only if their correlation is less than 1/3 in absolute value. (See the appendix.) It can be shown that this implies that the correlation between u_{1i} and u_{2i} (the usual linear probability dichotomous disturbances) must be less than 1/4 in absolute value. In fact, the latter correlation is bounded by $\sqrt{p_{1i}q_{1i}p_{2i}q_{2i}}$, where p_{1i} = Prob $(y_{1i} = 1)$, $q_{1i} = 1 - p_{1i}$, and p_{2i} and q_{2i} are defined analogously. In practice, correlations between dichotomous variables tend to be low, and perhaps most data will meet these conditions. In any event, minor violations lead to undefined values of the joint distribution in just a small fraction of the observations.

1977). Two kinds of errors can occur: harmless defendants may be unnecessarily detained, and dangerous or flight-prone defendants may be released. The first kind of error, unnecessary confinement, is of genuine importance—not only to society, but even more so to the defendants. On any given day, tens of thousands of Americans are being detained. On March 15, 1970, for example, more than 80,000 people were being held before trial in American jails (Thomas, 1976, p. 31). Many such people would do no harm if released; many others are not convicted. Unnecessarily maintaining them in custody is expensive. Apart from the government's cost of feeding, housing, and guarding them, wages are forgone, jobs are lost, and marriages are destroyed as detentions extend to weeks and months. The jails themselves are often overcrowded, with abysmal food, poor sanitation, and no exercise facilities. Insect and rodent infestations are common. So appalling are conditions in some jurisdictions that defendants have been known to plead guilty in hopes of escaping to the comparative pleasures of state prisons. For both the government and the detained individuals, then, jailing harmless defendants before trial is a costly error.

The second type of error in pretrial systems, releasing flight-prone or dangerous defendants, also entails substantial losses. If judges mistakenly release accused individuals who skip bail and flee, justice is thwarted. And if they discharge dangerous defendants before trial, innocent bystanders suffer. Indeed, when ordinary citizens talk about failures of the bail system, the latter sort of mistake is usually what they have in mind. Except for jail riots, the only event in the pretrial system that merits newspaper space is the horrible crime committed by a defendant recently released on bail or recognizance.

The cost of crime committed by released defendants often goes unmentioned in evaluation studies of pretrial release, however. The standard justification for its omission is that the Constitution prevents jailing people for crimes they have not yet committed. They are innocent until proven guilty. Hence "dangerousness" is not a permissible consideration in bail decisions. However unfortunate pretrial crime may be, the argument goes, the Constitution requires that it be ignored, both by judges and by policy analysts. In consequence, bail law in many states excludes dangerousness as a criterion in pretrial release decisions. Judges are instructed to consider only risk of flight and to assess that risk by examining the defendant's ties to the community. In the same spirit, bail administration agencies in some jurisdictions recommend defendants for release by assessing community ties, either informally or by using a formal point scale that ignores dangerousness (Kirby, 1977, app.).

In practice, however, judges certainly take dangerousness into account, and a growing number of states permit them to do so legally. On this view, once the police satisfy a judge that enough evidence exists to justify a trial, the defendant no longer has all the rights and protections of a citizen against whom no substantiated criminal charges have been brought. Thus detention before trial has quite a different character from arresting otherwise innocent citizens on suspicion of imminent criminal acts. The ordinary citizen enjoys constitutional protection from preventive detention; the dangerous defendant need not. Conflating these two groups, it is said, makes it impossible for a pretrial system to function at all. Every dangerous defendant would have to be released before trial—no matter how serious the charge, how overwhelming the evidence, or how likely a repetition of the crime. Presumably, no one wants a repeat of the recommendation, by the New York agency responsible for evaluating defendants, that David Berkowitz, who was accused of being the notorious multiple murderer Son of Sam, be released on his own recognizance.[5]

Evaluating a pretrial release system, then, cannot be done without deciding which costs may legitimately be considered in the evaluation. Doing so requires an ethical or normative judgment that empirical methods cannot supply. Whichever decision is made, however, a policy analyst will wish to know whether judges release the appropriate defendants (as judged by their probability of fleeing the jurisdiction or committing another crime while on release). To do so, the analyst must be able to compare the actual behavior of the released group with the expected behavior of the unreleased group had they been released. But the only data available for estimating the skip rate

---

[5]In fact the recommendation was a clerical error, since defendants with the most serious charges are excluded from favorable recommendations. Apart from the charge, though, Berkowitz had enough points for a recommendation of release: his ties to the community were impeccable.

or dangerousness of those detained come from the group of defendants released. This group is far from randomly selected: judges are compelled both by legal norms and by political pressure to release only the safer and more trustworthy individuals. Other defendants simply remain in jail. All the evidence on rearrest and flight during pretrial release comes from a selected sample—namely, those defendants who were expected to do well. Thus the evaluation of a pretrial release system depends on data that exhibit all the properties of nonrandomly selected observations.

Previous studies have evaluated release systems as though the observations met the textbook definition of random draws from a population. In consequence, the literature is replete with statistical anomalies of the sort predicted by Chapter 4: low R^2, poor predictions, and implausible signs on regression coefficients. For example, Landes (1974b, p. 313) finds that with certain other factors controlled, accused felons are less likely to be rearrested while on bail than are accused misdemeanants. Although it is not impossible to concoct hypotheses that account for this finding, common sense suggests that just the reverse is true: felony defendants are more hardened cases with a greater likelihood of recidivism. In any event, the nature of the sample makes it impossible to learn the truth with elementary statistical methods. In nonrandomly selected samples like those produced by a pretrial release system, regardless of whether felony defendants are more or less likely to be rearrested, they will appear less dangerous on release than they really are.

To understand why this is so, consider a stylized release system in which judges release just two classes of defendants—those accused of misdemeanors and those felony defendants who were injured in the course of their alleged crime and are now confined to wheelchairs. Other felony defendants are held under bail amounts so high that they are never able to meet them. Now a policy analyst draws a sample from the population of defendants released under this system. It consists solely of accused misdemeanants and accused felons confined to wheelchairs. By assumption, the latter group is unable to commit another crime. Their rearrest rate is 0 percent. The misdemeanant defendants, on the other hand, are rearrested at a 15 percent rate. Hence when the analyst investigates the nonrandom sample of releasees, he or she will find that *in the sample* felons skip 15 percent less often. But of course this finding implies nothing whatever about the behavior of felony defendants who are not disabled and therefore were never released. Were they discharged, their rearrest rate might be double the accused misdemeanant rate. Conventional statistical techniques applied to the available sample do not allow one to infer anything at all about the probable behavior of defendants in detention.

Actual pretrial release systems do not choose defendants for release in the stark manner of the stylized system. Even so, the same statistical anomalies occur. More serious charges and prior criminal records prejudice judges against releasing a defendant. Those who obtain release over those initial objections are chosen because their risk of flight and rearrest is relatively low. Thus the sample of released defendants with serious charges or extensive criminal records will inevitably be biased toward good risks. In consequence, the nature of the charge and the defendant's criminal record may predict poorly or in a counterintuitive direction (see, for example, Angel et al., 1971). Reformers have usually concluded that pretrial crime cannot be predicted. But poor predictive power in the released group does not imply either that rearrest is unpredictable in the full sample of defendants or that judges should alter their decisions. It is very likely that the unreleased defendants would have done less well than those released. As Chapter 4 showed, low R^2 and reversed OLS coefficients in a selected sample are perfectly consistent with appropriate decisionmaking and high predictive power.

An accurate evaluation of the performance of a pretrial release system requires a correction for selection bias. In turn, this means that the analyst must model both the process by which defendants are selected for release and also the mechanism that determines their behavior while on release. For that purpose, a study is reported here that uses the 1975 Washington, D.C., Bail Agency sample discussed in Chapter 3. Since this data set is considerably larger than most others used in pretrial studies, more sophisticated statistical analysis is feasible even with the noisy data typical of criminal justice systems.

The first step in the analysis was to specify the selection equation; the dependent variable must be some version of release before trial. The underlying behavioral process here is quite complex. As noted above, defendants can gain their freedom in two quite different ways—either on their own recognizance or by posting bond. They can post bond in several different ways—by raising their own funds, by borrowing them from relatives or friends, by convincing a bondsman that they are worth the risk of accepting a 10 percent payment in lieu of the full amount, or, finally, by convincing the judge that they should be allowed to post the same 10 percent with the court ("cash bail"). Modeling all this complexity is not feasible, especially since bail amounts are not recorded in the data. To simplify the analysis, therefore, release was defined as "release on own recognizance (at the initial bail hearing)." All other defendants, including those who gained release on bail at the initial hearing, are treated initially as not released. This step eliminates the need to model the complex process of posting bail. In particular, the many routes to release via bond posting are irrelevant; the selection equation need describe only the judge's decision to release on own recognizance. The sole cost is a one-sixth reduction in the sample of defendants on release—a quite tolerable loss in a sample of this size.

In the behavioral equation, two dependent variables are of interest. The first, failure to appear for trial (or "skip"), is the least problematic legally but the most troublesome empirically. Everyone agrees that defendants who are likely to flee may be detained without offense to the Constitution. The problem lies in determining what constitutes flight. Many defendants, especially in lengthy trials with many appearances, oversleep or simply forget one or more appearances. Legally, failure to keep a court appointment constitutes skipping, and bench warrants are issued each time. In practice, however, defendants usually return voluntarily or are located by the Bail Agency or the police warrant enforcement squad. The bench warrants are then voided with no more penalty to the defendant than a lecture. Thus meaningful skips are probably best defined as those instances in which the defendant fails to appear and continues to avoid contact with the courts. In that case, a failure to appear becomes the final disposition of the case.

Even this definition of failure to appear has its difficulties. As several studies have shown, the skip rate reflects not only the characteristics of defendants but also the effort put into reminding defendants of their trial dates and finding them when they fail to come to court. In Washington, for example, failure to appear rose 22 percent after funds for notification of defendants were eliminated in 1976 (District of Columbia Bail Agency, 1977, p. 20). Moreover, not all skips are troublesome. Suppose that a group of prostitutes is arrested because there are too many of them in front of a good hotel one evening. If they post $300 bond each and forfeit it by fleeing the jurisdiction, approximate justice has been done.

It follows that defining and predicting skip is not easy and that evaluating judges by their skillfulness at reducing the skip rate may not always be sensible. Other agencies of the pretrial release system may be more responsible for pretrial flight, and no one may care to eliminate skip in every case anyway. Thus even with corrections for selection bias, skip may be difficult to predict statistically—yielding poor fits, unstable coefficients, and results that tell the analyst more about the priorities of judges and police than about the characteristics of defendants. In fact, several preliminary studies with the 1975 Washington data showed very little success in predicting skip, even with corrections for selection bias. For that reason, skip was eliminated as a dependent variable in the behavior equation.

The other indicator of performance on pretrial release is rearrest. How well are judges performing in minimizing pretrial crime? By contrast with skipping, the use of rearrest as a criterion is controversial. But the fact of rearrest is unambiguous, clearly undesirable nearly all the time, and not subject to much improvement by better management at the Bail Agency. It is also of far greater political importance than failure to appear. For all these reasons, rearrest while on pretrial release is a useful measure of judicial failure. Rearrest therefore was made the outcome variable.

The first step in the estimation procedure was to define the sample. In order to trace defendants through their pretrial period, only those defendants arrested from January to June 1975 were included. Since few trials continued beyond six months, virtually all rearrests could be found within the 1975 year for which data were available.

The next step was to find a suitable equation for forecasting release on recognizance. The complete set of exogenous variables used in Chapter 3 was examined—first in bivariate fits to the release variable and then in multiple regressions. Independent variables were retained if they had plausible substantive interpretations and estimated effects of 2 percent or more across the alternative specifications. Thus a variable such as narcotics use, with a statistically significant effect on the probability of release, was eliminated because its effect was almost always below 2 percent, whereas a statistically insignificant variable such as being accused of a crime against property was retained, since the best estimate of its effect was 7 percent and it was quite likely that the true value exceeded 2 percent. These rules leave the researcher less choice than might be imagined: most of the variables in the equations were entered initially because common sense, courtroom observation, and the literature suggest that they belong there. Examples include the severity of the charge, prior record of the defendant, and his recent criminal behavior. The Bail Agency interviews defendants and recommends (on a subjective basis, not a formal point scale) whether they should be released on recognizance; that variable too was entered. Each of these variables proved to have substantial and statistically significant effects. Other characteristics of the defendant and the charge (for instance, ties to the community such as residence in the District of Columbia) were simply tested for importance and the consistently influential ones were retained. The latter characteristics are essentially control factors whose inclusion or exclusion made little difference in the other coefficients. All variables used are defined in Table 4.

TABLE 4. VARIABLE DEFINITIONS.

Explained (Left-Hand-Side Endogenous) Variables		
Released on Own Recognizance	1	If defendant was released on own recognizance
	0	Otherwise
Rearrest	1	If defendant was rearrested while on release before trial
	0	Otherwise
Explanatory (Included) Exogenous Variables		
BAIL AGENCY RECOMMENDATION		
Recommended	1	If defendant was recommended by Bail Agency for release on own recognizance
	0	Otherwise

DEFENDANT'S PERSONAL CHARACTERISTICS

Variable	Value	Definition
Prior Convictions		Number of prior convictions in D.C. police records, excluding traffic offenses and D.C. municipal violations (8 = 8 or more)
Prior Skips		Number of times defendant has failed to appear for previous court proceedings, excluding traffic and municipal violations (8 = 8 or more)
Concurrent Address	1	If defendant considers self to have more than one permanent address (e.g., parents' and girlfriend's homes)
	0	Otherwise
Age		Defendant's self-reported age in years
Employed Off and On	1	If defendant reports being employed on irregular basis
	0	Otherwise
Alcohol	1	If defendant admits to being an alcoholic or having been treated for alcoholism
	0	Otherwise
Narcotics	1	If defendant admits prior or current drug addiction or treatment
	0	Otherwise
Bond Status	1	If defendant was on pretrial release for another crime when arrested
	0	Otherwise
Under Sentence	1	If defendant is on probation, parole, work release, or diversion from another sentence
	0	Otherwise

NATURE OF THE CHARGE

Variable	Code	Definition
Dangerous Offense	1	If most serious charge is "dangerous" under sec. 23-1331 (4) of the Bail Reform Act (essentially either "violent" or drug-related)
	0	Otherwise
Crimes Against Person	1	If most serious charge is crime against person (all categories of homicide, kidnapping, assault, sexual assault, robbery, child cruelty)
	0	Otherwise
Crimes Against Property	1	If crimes against property (all categories of arson, burglary, larceny, extortion, fraud, embezzlement, forgery, stolen property, stolen vehicles, and miscellaneous, including obstruction of mail and cruelty to animals)
	0	Otherwise
Crime Severity		A Bail Agency severity code for the gravity of the most serious charge as reflected in maximum and minimum sentences prescribed by law. Range: 1–135, where 1 = first-degree murder and 135 = soliciting for prostitution (Welsh and Viets, 1976, pp. D1–D8)
Felony	1	If most serious charge is a felony
	0	Otherwise
Morals Offense	1	If most serious charge relates to morals (dangerous drugs, gambling, sex offenses)
	0	Otherwise

Public Order Offense	1	If most serious charge relates to public order (weapons, obstruction of justice, bribery, flight and escape, parole/probation violation, rioting, possession of crime tools, harboring a fugitive, and introducing contraband into a penal institution)
	0	Otherwise
	DEFENDANT'S RESOURCES	
Income		Defendant's self-reported legitimate income (0 = none or missing; 1 = \$0–80/week, 2 = \$80–100/week, 3 = \$100–\$120/week, 4 = \$120–160/week, 5 = \$160–200/week, 6 = \$200–300/week, 7 = \$300–360/week, 8 = more than \$360/week)

Variable definitions for Tables 5 and 6. The data come from the 1975 defendant survey by the Washington, D.C., Bail Agency (Welsh and Viets, 1976).

Both race and gender were excluded from the equation. Being male had no detectable influence; blacks were estimated to have perhaps a 2 percent disadvantage, never quite attaining statistical significance. Whether the racial difference is pure discrimination, the effect of excluded factors, or statistical artifact cannot be determined with the available data. In any event, while these variables are important for certain purposes (studying whether judges are fair), they should be excluded as variables when an analyst assesses the statistical predictability of rearrest or compares judicial decisions to a theoretical optimum. Judges are constitutionally barred from using race or gender information in making their own forecasts; by the same token, a forecast equation for predicting the dangerousness of defendants would surely be unconstitutional if it made use of racial or gender variables. Thus both variables would have been dropped for the

TABLE 5. PREDICTION EQUATIONS FOR PROBABILITY THAT A DEFENDANT IS RELEASED ON OWN RECOGNIZANCE—COEFFICIENTS (STANDARD ERRORS).

Variable	GLS Estimates
Recommended	0.418** (0.014)
Prior Convictions	−0.053** (0.006)
(Prior Convictions)^2	0.043** (0.0008)
Prior Skips	−0.071** (0.013)
(Prior Skips)^2	0.017** (0.004)
Concurrent Address	−0.030* (0.015)
Age	0.0012** (0.0003)
Alcohol	−0.050** (0.016)
Bond Status	−0.139** (0.012)
Under Sentence	−0.076** (0.011)
Crime Against Person	−0.099* (0.047)
Crime Against Property	−0.071 (0.046)
Crime Severity	0.0040** (0.0005)
(Crime Severity)^2	−0.000031** (0.000003)
Felony	−0.139** (0.018)
Crime Against Morals	−0.109* (0.046)
Crime Against Public Order	−0.146** (0.047)
Income	0.0051** (0.0013)
Intercept	0.530** (0.054)
N	7,234
R^2 (initial OLS round)	0.41

* Statistically significant (two-sided) at 0.05.
** Statistically significant (two-sided) at 0.01.

Linear probability GLS estimates (corrected for heteroskedasticity) for the probability that a judge will release the defendant before trial on own recognizance. Data are taken from the 1975 District of Columbia Bail Agency sample (Welsh and Viets, 1976).

purposes of this study in any case; happily, the statistical evidence for their inclusion is very weak as well.

The final version of the selection equation is given in Table 5. Since the dependent variable, Released on Own Recognizance, is dichotomous, the equation is of the linear probability type. (See Equation (3) above.) The coefficients and standard errors are from the second-stage GLS estimates, with heteroskedasticity corrected. The coefficients tell the expected story: the Bail Agency's recommendation is extremely helpful in gaining release on OR, and income also seems to persuade judges that a defendant is trustworthy. On the other hand, prior convictions, prior skips, recent criminal activity, and more serious charges are harmful to release prospects. The negative coefficient on alcohol use is due partly to judges' diverting defendants to a detoxification center. All these variables exhibit effects very likely to be substantively significant, and all but one are statistically significant at 0.05 as well, with most significant beyond 0.01. Overall, the (first-stage OLS) fit is very good for a dichotomous prediction ($R^2 = 0.41$).

In the outcome equation, the dependent variable, Rearrest, is dichotomous. Its equation is specified as a linear probability model, just as the treatment variable, Released on Own Recognizance, was. The corresponding selection bias estimator is applied in the outcome equation. (See Equation (4) above.) As in the selection equation, most variables were entered into the outcome equation because they were almost certain to be consequential in predicting additional crime. Examples include narcotics use, prior record, bond status, and being under sentence. More severe charges, irregular employment, and prior skips also mattered, probably because they indicate a more criminal life-style. Finally, the recommendation of the Bail Agency was entered as a test of its usefulness, since judges rely on it so heavily. As before, race and sex were deliberately omitted; neither they nor any other demographic factors or crime categories had consistent effects across alternative specifications.

The first column of Table 6 gives the results of the selection bias correction. Once again the control variables show the expected effects. Prior convictions and skips, recent crime, narcotics use, and more serious charges all make rearrest more likely. The effects are all substantial and statistically significant, most beyond the 0.01 level. Essentially, past crime and drug use predict future crime, while other personal characteristics do not (compare Holt, 1976).

TABLE 6. PREDICTION EQUATIONS FOR PROBABILITY THAT A DEFENDANT ON PRETRIAL RELEASE WILL BE REARRESTED—COEFFICIENTS (STANDARD ERRORS).

Variable	Selection Bias Corrected	OLS (OR Release Only)	OLS (All Releases)
Recommended	−0.037	−0.007	0.014
	(0.041)	(0.016)	(0.013)
Prior Convictions	0.025**	0.022**	0.016*
	(0.0060)	(0.004)	(0.004)
Prior Skips	0.092**	0.079**	0.033
	(0.034)	(0.026)	(0.021)
$(\text{Prior Skips})^2$	−0.018*	−0.016*	−0.00079
	(0.0072)	(0.007)	(0.00526)
Employed Off and On	0.058**	0.056**	0.040*
	(0.022)	(0.019)	(0.017)
Narcotics	0.088**	0.084**	0.088**
	(0.017)	(0.014)	(0.012)
Bond Status	0.285**	0.266**	0.223**
	(0.032)	(0.017)	(0.014)
Under Sentence	0.044*	0.038*	0.043**
	(0.023)	(0.017)	(0.014)
Dangerous Offense	0.063**	0.056*	0.036
	(0.024)	(0.023)	(0.021)
Crime Severity	−0.0021*	−0.0019*	−0.0019*
	(0.00088)	(0.0008)	(0.0007)
$(\text{Crime Severity})^2$	0.000022**	0.000020**	0.000019**
	(0.000006)	(0.000005)	(0.000004)
$p_{2i}q_{2i}\hat{u}_{1i}$	−0.443		
	(0.444)		
Intercept	0.119*	0.082*	0.071*
	(0.059)	(0.033)	(0.028)
N	4,973	4,973	5,998
R^2	0.113	0.112	0.098

*Statistically significant (two-sided) at 0.05.
**Statistically significant (two-sided) at 0.01.

Three linear probability models for the probability that a defendant on pretrial arrest will be arrested. The observations are all defendants released during January–June 1975 in Washington, D.C., and for whom the Bail Agency had full information (Welsh and Viets, 1976). Defendants released on bail are excluded from the first two equations. The correct standard errors for the selection-bias corrected coefficients, which are presented here, average about 20 percent above the nominal OLS values.

The only variable with uncertain effects is precisely the one on which the judges rely so heavily—the Bail Agency's recommendation. The best estimate is that, all else equal, receiving a positive recommendation from the agency lowers the probability of rearrest by less than 4 percent and almost surely less than 10 percent. Indeed, zero is a considerably more likely value than 10 percent. Moreover, across other specifications the estimated effect of the recommendation never exceeded 4 percent; most estimates fell in the 2 to 4 percent range and the rest lower. (The recommendation was no better in predicting skip. In a small number of attempts to predict that variable, the typical effect was 2 percent.) It seems clear that the Bail Agency recommendation does not justify the faith judges have in it.

It is also noteworthy that the coefficient on the nonlinear term $p_{2i}q_{2i}\hat{u}_{1i}$ is negative, meaning that defendants chosen by the judges for release in spite of being relatively poor risks (as measured by the variables in the selection equation) perform better on release than their objective characteristics would indicate. That is, the judges apparently have access to information about defendants that is not captured by the measured variables in the selection equation (courtroom demeanor, for example). The impact of the additional information is modest but not trivial. For a somewhat risky defendant whose estimated probability of release is 0.5 and whose probability of being rearrested (as forecast by the measured factors) is 0.3, the estimated reduction in expected probability of rearrest due to the judge's release decision would be approximately 5 percent. Thus the defendant's estimated probability of rearrest would be 25 percent rather than 30 percent. In a sample where the average rearrest probability is 18 percent, reductions of five percentage points in rearrest probabilities are not negligible.

The judges also weight the measured variables quite well. Table 5 shows that judges are greatly concerned about the severity of the defendant's charge, his bond status, his being under sentence, and his prior convictions and skips. These are precisely the aspects of prior behavior that predict rearrest in Table 6. Although the coefficients in the two tables relate to different dependent variables (OR Release versus Rearrest) and thus are not directly comparable, corresponding coefficients should be approximately proportionate (though of opposite sign) if judges are successfully choosing releasees to minimize rearrest rates. As inspection of the two tables shows, the coefficients do exhibit this kind of relationship to a remarkable degree. That is, most of the key variables are appropriately used by the judges in an effort to minimize pretrial crime. (See Landes, 1974b, for similar findings.)

The differences between the two equations reduce to three: first, as noted above, the judges rely far too much on the Bail Agency's recommendation; second, the judges slightly overestimate the importance of alcoholism history and underestimate that of drugs; and third, judges pay attention to the nature of the charge in ways that do not predict the incidence of pretrial crime. The third difference probably reflects the judges' legal obligation to take account of the current charge. That is, higher probabilities of rearrest are required to hold defendants accused of minor crimes. Thus the importance of the current charge in the judges' thinking probably results from the legal system and norms of justice, rather than from errors of decision-making. In short, then, the only major difference between judicial decisions and rules that would minimize rearrest is that the judges place too much faith in the Bail Agency's recommendation.

The reliability of the conclusions from the selection bias correction was assessed in a variety of ways. First, the forecasts of the behavior equation were computed to ascertain whether they fell within the meaningful interval for probabilities, namely 0 to 1. All did. The forecasts were also grouped into intervals of 0.1 and compared with the actual values. No anomalies were found, and the linearity of the fit was quite good.

A variety of specifications were tried for the effect of the residual $\hat{u}_{1i}$. First the term $p_{2i}\hat{u}_{1i}$ was added to the equation; then the latter was replaced by the term $q_{2i}\hat{u}_{1i}$. Both these new specifications allow the bivariate distribution of the errors in the selection and outcome equations to be somewhat different from that previously assumed. In each case, however, the result was virtually identical coefficients on all the original variables, no substantial improvement in the fit, and slightly tighter confidence intervals. In particular, the coefficient on the original selection bias correction term $p_{2i}q_{2i}\hat{u}_{1i}$ remained stable and became significant at 0.05 in one case and just missed significance in the other. The only aspect of the regression that worsened was the coefficient on the recommendation of the Bail Agency, which became quite noisy.[6] In short, no other specification altered the substantive coefficients more than slightly, and all the alternatives confirmed the impact of the residual from the selection equation.

The outcome equation was also re-estimated on an independent sample. The equation was applied to the cases originating between July and December in Washington. As noted earlier, rearrest information is incomplete for these defendants, so that the test must be interpreted with caution. Since many rearrests undoubtedly occurred after December 31 and went unrecorded, this sample has a lower rearrest rate than the January to June group. (14 percent rather than 18 percent). Thus if the equation is correctly specified, the coefficients should be similar or slightly smaller in the new sample, and in general they were. The impact of bond status remained at 29 percent, the effect of being under sentence fell from 9 percent to 7 percent, the coefficient on narcotics use dropped from 9 percent to 6 percent, and the effect of being under sentence rose from 4 percent to 7 percent. Most other coefficients were slightly lower than their values in the January–June sample, including that of the Bail Agency's recommendation, which fell to exactly 1 percent. The only anomaly was the coefficient on the residual, which was positive (0.6 with a t ratio of 1.2). The latter result, if not simply sampling error, is probably a statistical anomaly due to the nature of this sample.[7]

Finally, a more substantive check was tried. One argument against the use of rearrest as an evaluation criterion is that the police may simply "round up the usual suspects." That is, once the defendant is known to the police on the original charge, his subsequent arrest record may be influenced less by his behavior than by his notoriety.

---

[6]The cause of the large standard error is the multicollinearity of the Bail Agency's recommendation with the residual from the release equation. Recall that the recommendation is the key component of predicted release on recognizance. Hence unrecommended defendants who gain release will have large positive residuals, making those residuals negatively correlated with the recommendation.

[7]The justice system does very little business in December, presumably because only the worst cases are arrested—and even among them it appears that many are freed on recognizance in the spirit of the season. If this is true, in this sample the largest residuals from the selection equation would correspond to relatively dangerous December defendants. Their repeat crimes in December are less likely to be charged, however, and those in January and later are not recorded in this data set. Hence the largest residuals are associated with few recorded rearrests, inducing a positive coefficient on the residual.

When judges are highly professional, as they are in Washington, this argument may be checked straightforwardly by substituting *conviction* for arrest as the measure of behavior on release. If the police are padding their clearance rates with bad arrests, the coefficients should show a very different pattern with the change in dependent variable. In the extreme case, where the police simply rearrest previous defendants whenever the crimes are similar, the factors predicting rearrest would also predict a *failure* to convict. By contrast, the actual results showed the pattern one would expect if rearrest is legitimate: all coefficients were reduced approximately proportionately from the original values (reflecting the fact that not all rearrests lead to conviction). Thus the effect of bond status falls from 29 percent to 17 percent and that of narcotics from 9 percent to 5 percent, while being under sentence has its effect reduced from 4 percent to 3 percent and that of the Bail Agency's recommendation drops from 4 percent to 1 percent. The coefficient on the residual term also falls from –0.4 to –0.2. In summary, the results support the view that rearrest is an independent decision by the police rather than a simple corollary of the first arrest. In turn, that finding confirms the good predictive performance by the judges in minimizing rearrest.

By way of comparison, Table 6 also shows the coefficients obtained from ordinary regression for the identical sample (OR releases only). Since the group released on their own recognizance were good bets for release (almost 90 percent of them were recommended by the Bail Agency), their residuals in the selection equation are small. Hence adding these residuals to the outcome equation to correct for selection makes relatively little difference. Consequently, ordinary least squares applied to the OR releases only should generally have coefficient estimates not very different from the selection bias corrections; Table 6 shows that this expectation is met.

Table 6 also shows the OLS results when all defendants released at their initial hearing (OR plus bail) are included. These findings correspond to what the typical analysis found in the pretrial release literature might produce: everyone on release enters the sample and OLS is applied. The sample differs from the first OLS sample in that 17 percent of it consists of defendants who were refused release on recognizance but made bail at their first hearing. These defendants are fairly heavily selected from among all defendants assigned bail, since many defendants assigned bail never post it, and most who do so post it later than the first hearing. With these additional observations in the sample, then, the classic signs of selection bias should make their appearance. Indeed, despite the greater diversity of the sample, R^2 is down and coefficients are generally lower, especially on those variables that weigh heavily in the judges' decisions to release. In particular, the estimated effect of the Bail Agency's recommendation is slightly positive (though not statistically significant), meaning that recommended defendants are more likely to be rearrested. This is the same sort of sign reversal discussed in the hypothetical case of the felony defendants in wheelchairs; unrecommended defendants who get out anyway are such good risks that they do better than the average recommended defendant, all else being equal. In short, a conventional OLS analysis of this sample produces the same sort of anomalous findings typical of the rest of the pretrial release literature.

In summary, then, correcting for selection bias gives substantially, more meaningful results than a conventional OLS analysis. Even when the coefficients from the selection bias correction and from the OLS computations are not strikingly different, however, correcting for selection bias remains important. For the OLS methods differ very substantially from the bias-corrected estimates in their forecasts. This fact is of key importance in pretrial release studies because so much turns on the probable performance on release of those defendants who were remanded to jail. If pretrial decisions, for all their facade of legal rationality, turn out to be essentially arbitrary, the jailed group should be no worse risks than those on release. And if that is true, the case for near-universal release before trial becomes very strong. Thus obtaining a dependable statistical forecast for the unreleased group is critical to assessing how well a pretrial release system is performing.

The principal difference between the selection bias forecasts and their OLS counterparts is that the former take account of the information in the disturbance term from the first equation. Forecasts may be dramatically altered as a result. In the study of rearrest, for example, the first equation's residuals represent what the judges know about defendants that is otherwise unmeasured. The OLS forecast simply ignores this private judicial information. The result is that without corrections for selection bias, the efficiency of the judiciary in pretrial decisions is substantially underestimated.

To see this effect, consider the *released* group in the Washington sample. For them, forecasts can be obtained in the usual way from Equation (4). That is, after obtaining the appropriate coefficient estimates, including the coefficient on the nonlinear term that includes the first-equation disturbance, one inserts the right-hand-side variables into the original estimating equation, sets the disturbance term v_{2i} to zero, and computes the result. Apart from the addition of the nonlinear term, this forecast is no different from that of any other regression forecast.

For the group not released on their own recognizance, the appropriate forecast is obtained somewhat differently, since the correct form of the residual from the selection equation is not $\hat{u}_{1i}$, as with the released group, but rather $-(1 - \hat{u}_{1i})$. Substituting the latter quantity in place of $\hat{u}_{1i}$ and then carrying out the forecast in the same way as for the released group gives the forecast. That is, the correct forecast for the nonselected group is

$$\hat{y}_{2i} = \hat{p}_{2i} + \hat{b}_{k+1,2}\hat{p}_{2i}\hat{q}_{2i}(\hat{u}_{1i} - 1) \tag{5}$$

where $\hat{y}_{2i}$ is the forecast, $\hat{p}_{2i} = \hat{a}_2 + \hat{b}_{12}x_{1i} + \cdots + \hat{b}_{K2}x_{Ki}$, $\hat{a}_2$ and the $\hat{b}_{j2}$ are coefficient estimates from the nonlinear least squares regression in Equation (4), $\hat{q}_{2i} = 1 - \hat{p}_{2i}$, and $\hat{u}_{1i}$ is the residual from the selection Equation (3).

With this machinery, the probable rearrest rate of those defendants who were not released on their own recognizance can be forecast. The forecasts with selection bias corrected, along with the OLS forecasts, are displayed in Table 7. All three estimators forecast crime in the OR release group quite well: since the coefficients were estimated using this data, the good fit is statistically inevitable.

The striking differences emerge in the forecasts for the group not released on OR at the initial hearing. All the estimates agree that the jailed group is considerably more risky than the group released, but the estimates with selection bias corrected are nearly twice as large as those obtained from the conventional analysis (OLS on all released defendants). The selection-corrected forecast is that, had everyone not chosen for OR been released from jail at the initial hearing, an astonishing 50 percent of them would have been rearrested before the conclusion of their cases. This estimate is probably near the truth:

TABLE 7. FORECAST REARREST RATES.

	Defendants on OR Release	Defendants Not on OR Release
Selection-corrected estimator	17.7%	50.1%
OLS (OR release sample)	17.8%	33.3%
OLS (recognizance + bail sample)	16.5%	28.3%
Actual	17.8%	—
N	4,973	2,765

Forecast rearrest rates from the three estimators in Table 6. The data are the January–June cases from the 1975 District of Columbia Bail Agency sample (Welsh and Viets, 1976).

although most of this group never left jail,[8] fully 18 percent of the total were actually rearrested, implying an enormous rearrest rate if all had been released. Seen in this light, the judges' ability to segregate these defendants from those on release seems very skillful indeed.

In summary, then, a clear view of the pretrial process in Washington emerges from the analysis when selection bias is corrected. By a combination of own recognizance and low bails, judges release immediately about three-quarters of all defendants. Most of the remainder are detained most of the time. Judicial skill in partitioning the defendants into these two groups is such that the arrest rate is less than 20 percent among releasees, but it would be more than twice that among the detainees if they were released. Releasing the detained group would be very costly: total pretrial crime would be estimated to rise by more than 50 percent. (This increase results from a rise in the rearrest rate from 17.7 to 25.4 percent and a 29 percent increase in the total number of defendants on release.) In other words, judges

[8]In Washington in 1971, four years before the period of this study, more than 70 percent of those initially assigned bail were detained until detention (Thomas, 1976, pp. 45, 73). This statistic is consistent with courtroom observation in 1979, which showed that bail reductions for detained defendants were frequently requested by their attorneys but infrequently obtained.

achieve a substantial reduction in pretrial crime from what would be achieved by releasing everyone yet still allow more than three-quarters of defendants to go free before trial.[9]

Thus the evaluation of judicial pretrial decision-making that emerges here is quite different from the standard appraisal in the empirical literature on pretrial release. Using conventional regression techniques, researchers have concluded that judges rely on criteria that do not predict defendant behavior. If judges paid more attention to Bail Agency recommendations, it is said, better decisions would result. Correcting for selection bias shows that, at least in a highly professional system like that of Washington, just the reverse is true: the criminal record and nature of the charge that judges rely on predict rearrest reasonably well, whereas community ties and Bail Agency recommendations have predictive power neither for rearrest nor skipping. Apart from excessive reliance on the Bail Agency, judges do a creditable job of selecting defendants for release. Rearrest rates remain high (18 percent) at the 1975 release rate (78 percent), and one can argue whether releases should be reduced to lower pretrial crime or raised at the cost of more detention. That debate is primarily philosophical and political, not empirical. What can be said on the basis of the evidence is that—given the fraction of people released before trial—the choice of releasees was quite good. At any level of release, selecting defendants for OR release by Bail Agency recommendations or by a point scale based on community ties, as reformers have long suggested, would worsen the pretrail crime problem noticeably.[10]

---

[9]By contrast, if the conventional forecasts (based on OLS on all released defendants) are accurate, one must assume implausibly that those defendants who were not released at their initial hearing ultimately spent more than half of their pretrial time on release.

[10]Readers unfamiliar with correlation measures for dichotomous dependent variables may wonder how the apparently good judicial predictions are consistent with an R^2 of just 0.11 in the rearrest forecast equation. Judging from the low explained variance, rearrest appears not to be very predictable. The solution is twofold. First, the relevant multiple correlation is not that in the selected sample but that in the full sample of defendants for whom forecasts must be made. The rearrest equation has an R^2 of 0.14 in the full sample.

Second, and more important, small multiple correlations for dichotomous dependent variables are perfectly consistent with powerful prediction, as noted long ago by Taylor and Russell (1939). Consider, for example, a pretrial release system in which one-third of the defendants have a 40 percent probability of rearrest while the remainder have just a 10 percent chance. If everyone were released, the rearrest rate would average 20 percent. On the other hand, if judges can discriminate perfectly among these two groups of defendants and release just the low-risk group, the rearrest rate will fall by half and total pretrial crime will drop by two-thirds. Thus dividing defendants into a 10 percent group and a 40 percent group constitutes a great predictive success. But, as a simple calculation will show, the R^2 over the entire sample for the judges in this example is just 0.12.

Appendix

This appendix sets out the theory of correcting for selection bias when the selection equation is a linear probability model.

A. Continuous Outcome Variable

Consider the two-equation system:

$$y_1^* = X_1\beta_1 + u_1^* \tag{A-1}$$

$$y_2 = X_2\beta_2 + u_2 \tag{A-2}$$

Here y_1^* and y_2 are $n \times 1$ vectors of dependent variables, X_1 and X_2 are $n \times k_1$ and $n \times k_2$ matrices of independent variables, u_1^* and u_2 are $n \times 1$ vectors of disturbances, and β_1 and β_2 are coefficient vectors to be estimated.

The vector y_1^* is unobserved. However, y_1 is observed, and its ith element is related to the ith element of y_1^* by:

$$y_{1i} = \begin{cases} 1 & \text{if } y_{1i}^* \geq 0.5 \\ 0 & \text{otherwise} \end{cases}$$

The selection equation (A-2) is a censored regression model: the ith element of y_2 is observed only if $y_{1i} = 1$. By convention, it is assumed that the observed values constitute the first m elements of y_2. (Note that m is random.)

The observed data for the outcome equation (A-2) are written as:

$$y_{21} = X_{21}\beta_2 + u_{21} \tag{A-3}$$

where y_{21}, X_{21}, and u_{21} each have m rows.

It is assumed that:

(i) Each element of u_1^* (denoted u_{1i}^*) has a uniform distribution over $[-0.5, 0.5]$.

Thus Equation (A-1) is a linear probability model. If X_{1i} is the ith row of X_1, then the probability that $y_{1i} = 1$ is denoted $p_i = X_{1i}\beta_1$, while the probability that $y_{1i} = 0$ is $q_i = 1 - p_i$. Denote by u_{1i} the disturbance

$y_{1i} - p_i$, and let u_1 be the corresponding vector. It follows that if Σ is a diagonal matrix with elements $p_i q_i$, then conditional on X_1, X_2, we have $E(u_1) = 0$ and $E(u_1 u_1') = \Sigma$.

(ii) For all i, p_i, $q_i \leq 1 - \delta$ for some $\delta > 0$.

(iii) Let u_{2i} be the ith element of u_2 and ρ a constant. Then the pair (u_{1i}, u_{2i}) is independent of (u_{1j}, u_{2j}) for all i, j $(i \neq j)$. Moreover, conditional on X_1, X_2, for all i:

$$\begin{aligned} E(u_{2i}) &= 0 \\ E(u_{2i}^2) &= \sigma^2 \\ E(u_{2i} | u_{1i}^*) &= 2\rho u_{1i}^* \\ \mathrm{var}(u_{2i} | u_{1i}^*) &= \omega^2 \end{aligned}$$

Thus observations are independent, so that each element of u_1^* and u_2 is correlated only with its corresponding element in the other vector—no serial correlation, for example. Moreover, the regression of u_{2i} on u_{1i}^* is linear and constant with fixed residual variance ω^2. Since the variance of a uniform distribution over the unit interval is 1/12, it follows that $\omega^2 = \sigma^2 - 4\rho^2 \mathrm{var}(u_{1i}^*) = \sigma^2 - \rho^2/3$.

(iv) X_1 is of full rank.

(v) Let $\bar{Z}_2$ be the matrix whose ith row is $\bar{Z}_{2i} = [X_{2i} \quad q_i]$, and let P be an $n \times n$ diagonal matrix with diagonal elements p_i Then lim $X_1'X_1/n$, lim $X_1'\Sigma^{-1}X_1/n$, and lim $\bar{Z}_2'P\bar{Z}_2/n$ are constant positive definite matrices. (Since $q_1 = 1 - X_{1i}\beta_1$, this last assumption requires that at least one row of X_1 be excluded from X_2 to avoid collinearity in Z_2.[11])

(vi) The elements of X_1 and X_2 are uniformly bounded in absolute value.

Assumption (iii) implies that if y_{2i} and X_{2i} are the ith elements of y_2 and X_2, then:

---

[11]Strictly speaking, this postulate requires that at least one row of X_1 be excluded from X_2 unless X_1 neither contains an intercept term nor is collinear with an intercept. The latter cases are rare in practice.

$$\begin{aligned} E(y_{2i}|y_{1i} = 1) &= X_{2i}\beta_2 + E(u_{2i}|y_{1i} = 1) \\ &= X_{2i}\beta_2 + 2\rho E(u^*_{1i}|y^*_{1i} \geq 0.5) \\ &= X_{2i}\beta_2 + \rho q_i \end{aligned} \tag{A-4}$$

since $y^*_{1i} \geq 0.5$ implies $u^*_{1i} \geq 0.5 - p_i$, and so conditionally u^*_{1i} has a uniform distribution over $[0.5 - p_i, 0.5]$ and its expectation is $(1 - p_i)/2 = q_i/2$.

If the censored regression (A-3) is to be estimated, (A-4) shows that q_i must be added to the equation as an additional independent variable. To estimate q_i, set

$$\hat{\beta}_1 = (X_1'\hat{\Sigma}^{-1}X_1)^{-1}X_1'\hat{\Sigma}^{-1}y_1$$

where $\hat{\Sigma}$ is a diagonal matrix with ith diagonal element $\bar{p}_i\bar{q}_i$, $\bar{p}_i = X_{1i}\bar{\beta}_1$, $\bar{q}_i = 1 - \bar{p}_1$, and $\bar{\beta}_1 = (X_1'X_1)^{-1}X_1'y_1$. If necessary, $\bar{p}_i$ may be restricted to some interval in [0, 1], say [0.01, 0.99]. By the theorem in the appendix to Chapter 3, the generalized least squares estimate $\hat{\beta}_1$ is consistent with asymptotic variance $(X_1'\Sigma_1^{-1}X_i)^{-1}$. Hence a consistent estimate of q_i is $\hat{q}_i = 1 - X_{1i}\hat{\beta}_1$; when y_{2i} is observed, this is simply the residual in the other equation. Denote the vectors whose elements are q_i and $\hat{q}_i$ by q and $\hat{q}$, and denote the first m elements of the latter vectors by q_1 and $\hat{q}_1$.

Since the elements of y_{21} are precisely those m elements of y_2 for which $y_{1i} = 1$, it follows from (A-4) that if $v_1 = u_{21} - \rho q_1$, a disturbance term with mean zero, then:

$$\begin{aligned} y_{21} &= X_{21}\beta_2 + \rho q_1 + v_1 \\ &= X_{21}\beta_2 + \rho\hat{q}_1 + w_1 \end{aligned} \tag{A-5}$$

where $w_1 = \rho(q_1 - \hat{q}_1) + v_1$. From the definition of $\omega^2 = \sigma^2 - \rho^2/3$ given in Assumption (iii) and the fact that the variance of a uniform distribution truncated to have range p_i is $p_i^2/12$, it follows that the variance of the ith element of v_1 is:

$$\begin{aligned} \text{var}(v_i) &= \text{var}(u_{2i}|y_{1i} = 1) \\ &= \text{var}(u_{2i} - 2\rho u^*_{1i}|y_{1i} = 1) + \text{var}(2\rho u^*_{1i}|y_{1i} = 1) \\ &= \frac{\omega^2 + 4\rho^2 p_i^2}{12} \\ &= \frac{\sigma^2 - \rho^2(1 - p_i^2)}{3} \end{aligned} \tag{A-6}$$

In Equation (A-5), since each element of $(q_1 - \hat{q}_1)$ converges in probability to zero, the elements of w_1 converge in distribution to those of v_1, which have mean zero, have finite variances, and are uncorrelated in probability with X_{21} and $\hat{q}_1$. This conclusion suggests a least squares estimator.

Letting $Z_{21} = [X_{21} \quad \hat{q}_1]$ and $\delta_2' = [\beta_2' \quad \rho]$, define:

$$\delta_2 = (Z_{21}'Z_{21})^{-1}Z_{21}'y_{21} \tag{A-7}$$

Now from (A-5):

$$\frac{Z_{21}'y_{21}}{n} = \frac{Z_{21}'Z_{21}\delta_2}{n} + \frac{Z_{21}'w_1}{n} \tag{A-8}$$

Thus from (A-7) and (A-8):

$$\hat{\delta}_2 = \delta_2 + \frac{(Z_{21}'Z_{21}/n)^{-1}Z_{21}'w_1}{n} \tag{A-9}$$

To evaluate $\hat{\delta}_2$ for consistency, note first that although Z_{21} is stochastic because its column $\hat{q}_1$ is estimated, the elements of $\hat{q}_1$ are each the same continuous function of $X_{1i}\beta_1$ over a closed bounded interval, that X_{1i} is uniformly absolutely bounded, and that $\hat{\beta}_1$ is consistent. Hence by Lemma 3 of Chapter 3, $\hat{q}_1$ meets the conditions of Lemma 2 and in all cross-products in Equation (A-9), q_1 may be substituted for $\hat{q}_1$ (and hence $\overline{Z}_{2i}$ for Z_{2i}) in taking probability limits and computing asymptotic variances.

Thus observe that by Chebychev's theorem and Assumption (vi):

$$\begin{aligned} \text{plim}\,\frac{Z_{21}'Z_{21}}{n} &= \frac{\lim \Sigma p_i \overline{Z}_{2i}\overline{Z}_{2i}'}{n} \\ &= \frac{\lim \overline{Z}_2'P\overline{Z}_2}{n} \\ &= \frac{G}{n} \end{aligned} \tag{A-10}$$

say, a constant positive definite matrix by Assumption (v). Hence the probability limit of the inverse also exists and is constant. Moreover, by an argument like that used in the proof of the theorem in the appendix to Chapter 3, the inverse in (A-7) exists with arbitrarily high probability in large samples and thus so does the estimate $\hat{\delta}_2$.

Finally, let w be the vector whose first m elements (for which y_2 is observed) are w_1 and whose remaining elements (for which y_2 is not observed) are the random variables that would have occurred had y_{2i} been observed. Define v similarly. Thus

$$w = \rho(q - \hat{q}) + v$$

Note that $(q - \hat{q})$ and v are uncorrelated in probability.

This implies that:

$$\text{plim}\ \frac{Z'_{21}w_1}{n} = \frac{\text{plim}\ \rho Z'_2P(q - \hat{q})}{n} + \frac{\text{plim}\ Z'_2Pv}{n} = 0 \qquad \text{(A-11)}$$

by Lemma 2 of Chapter 3 and by the fact that the elements of v have mean zero, bounded finite variances, and Z_2 and P are absolutely bounded. Hence by Equations (A-9), (A-10), and (A-11), $\hat{\delta}_2$ is consistent. It can also be shown to be asymptotically normally distributed.

By Equations (A-9) and (A-10), asymptotically:[12]

$$\text{var}\,(\sqrt{n}\hat{\delta}_2) = G^{-1}\text{var}\left(\frac{Z'_{21}w_1}{\sqrt{n}}\right)G^{-1} \qquad \text{(A-12)}$$

and, again asymptotically:

$$\text{var}\left(\frac{Z'_{21}w_1}{\sqrt{n}}\right) = \rho_2E\left[\frac{Z'_{21}(q_1 - \hat{q}_1)(q_1 - \hat{q}_1)'Z_{21}}{n}\right] + E\left[\frac{Z'_{21}v_1v'_1Z_{21}}{n}\right]$$

the cross-product term going to zero by the consistency of $\hat{q}$ and the boundedness of Z_2. In the expectations of $(q_1 - \hat{q}_1)\ (q_1 - \hat{q}_1)'$ and $v_1v'_1$, diagonal terms occur with probability p_i and off-diagonal terms with probability p_ip_j. It follows that:

$$\text{var}\left(\frac{Z'_{21}w_1}{\sqrt{n}}\right) = \frac{\rho^2\overline{Z}'_2PX_1(X'_1\Sigma^{-1}X_1)^{-1}X'_1P\overline{Z}_2}{n}$$

---

[12]The asymptotic variance of an estimator is defined as the limit of the variance of the estimator to which it converges in probability. In the derivation of the variance, expectations refer to the latter estimator's distribution.

$$+\frac{\bar{Z}_2' P[\Lambda + \rho^2(I - P)D]\bar{Z}_2}{n} \tag{A-13}$$

$$= \frac{H}{n}$$

say, where $\Lambda = \sigma^2 I - \rho^2 (I - P^2)/3$ is a diagonal matrix whose diagonal elements are $\text{var}(v_i)$ and $D = \text{diag}\, X_1(X_1'\Sigma^{-1}X_1)^{-1}X_1'$.

Then, equations from (A-11) and (A-12):

$$\text{var}\,(\hat{\delta}_2) = G^{-1}HG^{-1} \tag{A-14}$$

The variance (A-14) may be consistently estimated as follows. Let X_{11} be the first m rows of X_1 and let $\hat{P}_1$ be a diagonal matrix of dimension m whose diagonal elements are $\hat{P}_i$; the forecasts from Equation (A-1) correspond to the observed elements of y_2. The regression coefficients (A-7) give the estimate $\hat{\rho}$; and if the residuals from this regression are denoted $\hat{v}_1$, set $\hat{\sigma}_2 = [\hat{v}_1'\hat{v}_1 + \hat{\rho}^2\Sigma(1 - \hat{p}_i^2/3]/m$ using (A-6). (The sum is taken over just those data points for which y_2 is observed.)

Now let:

$$\hat{G} = Z_{21}'Z_{21}$$
$$\hat{A} = Z_{21}'X_{11}(X_1'\hat{\Sigma}^{-1}X_1)^{-1}X_{11}'Z_{21}$$
$$\hat{B} = \frac{Z_{21}'(I - \hat{P}_1^2)Z_{21}}{3}$$

An estimate of (A-14) is then:

$$\hat{\sigma}^2\hat{G}^{-1} + \hat{\rho}^2\hat{G}^{-1}(\hat{A} - \hat{B})\hat{G}^{-1} \tag{A-15}$$

If the matrices $\hat{A}$ and $\hat{B}$ each converge to constant matrices when divided by n, then (A-15) is a consistent estimator. If just one step is used in estimating the linear probability model, then $(X_1'\hat{\Sigma}^{-1}\, X_1)^{-1}$ should be replaced by $(X_1'X_1)^{-1}\, X_1'\hat{\Sigma}X_1\, (X_1'X_1)^{-1}$ in $\hat{A}$.

The following theorem has been established.

Theorem. Suppose that the ith row of $y_2 = X_2\beta_2 + u_2$ is observed only when the corresponding element of the linear probability model $y_1 = X_1\beta_1 + u_1$ is 1. Define $\hat{q} = y_1 - X_1\hat{\beta}_1$, where $\hat{\beta}_1$ is the generalized least squares estimate. Let X_{21} be the observed rows of X_2, and denote the corresponding elements of $\hat{q}$ and y_2 by $\hat{q}_1$ and y_{21}. Set

$Z_{21} = [X_{21} \quad \hat{q}_1]$. Then under Assumptions (i) to (vi), $\hat{\delta}_2 = (Z'_{21}Z_{21})^{-1}Z_{21}y_{21}$ is consistent and has asymptotic variance given by Equation (A-14).

B. Dichotomous Outcome Variable

When $y_2 = X_2\beta_2 + u_2$ is also a linear probability model, assume that:

$$y^*_2 = X_2\beta_2 + u^*_2$$

where the elements of u^*_2 are uniformly distributed over $[-\frac{1}{2}, \frac{1}{2}]$ and $y_{2i} = 1$ if $y^*_{2i} \geq \frac{1}{2}$ and is zero otherwise. Assume further that the conditional distribution of u^*_{2i} is:

(vii) $$f(u^*_{2i}|u^*_{1i} = u_{10}) = 1 + 12\rho^* u_{10}u^*_{2i}$$

where ρ^* is the correlation between u^*_{1i} and u^*_{2i}. It is straightforward to show that assumption (vii) defines a proper density function for $|\rho^*| \leq \frac{1}{3}$.

Now let $p_{1i} = X_{1i}\beta_1$ and $q_{1i} = 1 - p_{1i}$, and define p_{2i} and q_{2i} similarly. Then:

Lemma

(a) $$\text{Prob}(u^*_{2i} \geq \tfrac{1}{2} - p_{2i}|u^*_{1i} = u_{10}) = \int_{1/2-p_{2i}}^{1/2} f(u^*_{2i}|u^*_{1i} = u_{10})du^*_{2i} = p_{2i} + 6\rho^* p_{2i}q_{2i}u_{10}$$

(b) $$\text{Prob}(u^*_{21} \geq \tfrac{1}{2} - p_{2i}|y_{1i} = 1) = p_{2i} + 3\rho^*\, p_{2i}q_{2i}q_{1i}$$

Proof: Elementary calculus.

Now since $\text{Prob}(y_{2i} = 1|y_{1i} = 1) = \text{Prob}(u^*_{2i} \geq \frac{1}{2} - p_{2i}|y_{1i} = 1)$, the lemma implies that, conditional on $y_{1i} = 1$,

$$\begin{aligned} y_{2i} &= X_{2i}\beta_2 + \rho p_{2i}q_{2i}q_{1i} + v_i \qquad (i = 1, \cdots, m) \\ &= X_{2i}\beta_2 + \rho p_{2i}q_{2i}\hat{q}_{1i} + w_i \end{aligned} \tag{A-16}$$

where $\rho = 3\rho^*$ and $w_i = \rho p_{2i}q_{2i}(q_i - \hat{q}_i) + v_i$ and v_i has mean zero and variance r_is_i, with $r_i = X_{2i}\beta_2 + \rho p_{2i}q_{2i}q_{1i}$ and $s_i = 1 - r_i$.

Let $Z_{2i} = [X_{2i} \quad \hat{q}_{1i}]$ and $\gamma_{2i} = [\beta'_2 \quad \rho p_{2i}q_{2i}]$. Define $\tilde{\delta}'_2 = [\tilde{\beta}'_2 \quad \tilde{\rho}]$ as the vector that minimizes

$$\sum_{i=1}^{m}(y_{2i} - Z_{2i}\gamma_{2i})^2 \tag{A-17}$$

That is, $\tilde{\delta}_2$ is just the nonlinear least squares estimate of (A-16).

The methods of Malinvaud (1970, chap. 9) may be used to show that with slight modification of the preceding assumptions this estimator is consistent. Its asymptotic variance is given by replacing Λ by Λ^* and $\bar{Z}_2$ by $\bar{Z}_2^*$ in Equations (A-10), (A-13), and (A-14), where Λ^* is a diagonal matrix whose ith diagonal element is $r_i s_i$ and $\bar{Z}_2^*$ has an ith row whose elements are the derivatives of $Z_{2i}\gamma_{2i}$ with respect to δ_2. That is,

$$\bar{Z}_{2i}^* = [(1 + \rho q_{1i} - 2\rho p_{2i} q_{1i})X_{2i} \quad p_{2i} q_{2i} q_{1i}]$$

Let Z_{2i}^* be the estimate of the first m rows of $\bar{Z}_2^*$ (the observed rows), obtained by substituting $\tilde{\rho}$, $\tilde{p}_{2i}$, and $\tilde{q}_{2i}$ for their population values. Define $\tilde{G}^*$ and $\tilde{A}^*$ by inserting Z_{2i}^* for Z_{2i} in the definitions of $\hat{G}$ and $\hat{A}$ above. Define $\tilde{\Lambda}_1^*$ as the m-dimensional diagonal matrix whose ith diagonal element is $\tilde{r}_i \tilde{s}_i$, estimated in the obvious way for the n available observations on y_2. Then the variance of this second estimator may be estimated by

$$\mathrm{var}(\tilde{\delta}_2) = \tilde{G}^{*-1}(\tilde{\rho}^2 \tilde{A}^* + \tilde{C}^*)\tilde{G}^{*-1}$$

where $\tilde{C}^* = Z^{*\prime}_{21}\tilde{\Lambda}_1^* Z_{21}^*$. If just one step has been used in estimating the linear probability model, $\tilde{A}^*$ should be adjusted in the same way as for the previous estimator.

The heteroskedasticity in the disturbances of this estimator can be corrected, as they can be in the previous estimator, but the form of the variance matrix for the coefficients shows that such estimates are not fully efficient. Essentially the sampling errors of the estimates in each of these methods depends not only on the heteroskedastic variances of the disturbances but also on the errors in the estimate of $\hat{q}_i$, and these errors do not generate a diagonal variance matrix. Hence fully efficient generalized least squares procedures require the inversion of an n-dimensional square matrix, which is not practical for moderate to large samples. A simpler route to efficient estimates is to use the values generated by the estimators as starting values for one round of Newton iteration in a numerical search for maximum-likelihood estimates. The result will be asymptotically efficient estimators (Rothenberg and Leenders, 1964).

6

Theoretical Extensions and Data Analysis

INTRODUCTION

Previous chapters have discussed two varieties of quasi-experiments. In the first class, data are available for both an experimental and a control group, but individuals are not randomly assigned to treatments. In the second, evidence exists only for a subset of the population of interest and that subset is selected nonrandomly. In both cases it was shown how statistical ingenuity could circumvent the inferential pitfalls when prior information in the form of exclusion restrictions was present. The appropriate procedures for the easiest cases were explained and illustrated.

In practical circumstances, however, both nonrandomized control groups and censored samples produce data in complex ways that are not well described by the elementary models of previous chapters. To cope with more elaborate nonexperimental designs, a researcher needs a statistical specification that accurately captures the process generating the observations. This chapter therefore takes up extensions of previous chapters. No brief survey can encompass the inexhaustible diversity of quasi-experimentation. The hope is rather that common variations on the fundamental designs will be covered and that analysts will gain insight into how models might be formulated to cope with their own special cases.

NONRANDOMIZED EXPERIMENTS WHEN OUTCOMES INFLUENCE TREATMENT

Chapter 3 discussed estimation for the canonical nonexperimental design, in which membership in the experimental group may influence the outcome of the treatment but not the reverse. This postulate is a natural one when subjects join the experimental group before the treatment effect is measured, so that they cannot be assigned to treatments on the basis of how the experiment came out. In other circumstances, however, just this sort of assignment occurs. Consider, for example, the problem of assessing whether proportional representation schemes for electing legislators lead to fractionalized party systems (see, for example, Rae, 1971). On the one hand, the treatment (proportional representation) may influence outcomes (fractionalization); on the other, a splintered party system often leads to demands that proportional representation be introduced, so that outcomes may affect treatment. Reciprocal causation is at work.

Models of this kind may be handled in the same way as those of Chapter 3, as any elementary econometrics text will demonstrate. With cross-national data on votes, legislative seats, and the nature of the electoral system, for example, the effect of proportional representation can be estimated like the treatment effects of quasi-experiments without reciprocal causation. One simply writes down the outcome equation and estimates it by two-stage least squares as set out in Chapter 3. So long as at least one exclusion restriction is satisfied—there is an exogenous factor that influences the choice of electoral system but not the division of legislative seats—2SLS produces dependable estimates under the usual conditions.

In general, suppose several variables reciprocally influence each other so that all of them are endogenous. In the equation of interest, one variable is regarded as caused by the others; for example, the outcome of an experiment might be thought to depend on several aspects of the treatment. When this equation is identified,[1] 2SLS is

---

[1]A linear simultaneous equation must meet certain restrictions if it is to be *identified*—that is, statistically estimable. The simplest cases were discussed in Chapter 3. In general, a necessary (but not sufficient) condition is that the number of coefficients in the equation be no more than the total number of exogenous variables in the entire model. Further elementary discussion of this topic may be found in Hanushek and Jackson (1977, pp. 250–65). Fisher (1966) remains useful at a more advanced level; Hsiao (1983) is a sophisticated recent summary.

carried out as follows. Suppose that the equation to be estimated is:

$$y_1 = a_1 + c_2y_2 + \cdots + c_gy_g + b_1x_1 + \cdots + b_kx_k + u_1$$

where y_1 is the endogenous variable being "explained" by the equation, y_2 to y_g are other included endogenous variables, the x_j's are exogenous variables, the a's, b's, and c's are coefficients to be estimated, and u_1 is the disturbance term. Then the two-stage method is as follows:

A. Regress each of y_2 through y_g on all the exogenous variables in the model. Call the resulting forecasts $\hat{y}_2, \ldots, \hat{y}_g$.

B. Estimate the following regression by ordinary least squares:

$$y_1 = a_1 + c_2\hat{y}_2 + \cdots + c_g\hat{y}_g + b_1x_1 + \cdots + b_kx_k + v_1$$

C. Correct the standard errors of the coefficients in the same way as explained in the discussion of 2SLS in Chapter 3.

The result is consistent coefficient estimates and the appropriate standard errors. In short, when more than one endogenous variable occurs on the right-hand side of an identified equation, one simply purges all of them and then carries out 2SLS in the same way as before.

NONRANDOMIZED EXPERIMENTS WITH NONLINEAR VARIABLES

Nonlinear outcome equations that are linear in parameters are also easily handled.[2] Chapter 3 pointed out that only nonlinearities in *endogenous* variables are problematical. The same chapter showed how two-stage least squares can be extended to one such case, in which the treatment was chosen via a probit equation. Other endogenous nonlinearities in a structural equation can be dealt with in essentially the same way. Suppose, for instance, that in an equation "explaining" y_1, an endogenous variable y_2 and an exogenous variable x_1 appear together on the right-hand side in an interactive term, x_1y_2. This specification might mean, for example, that the impact of y_2 takes

---

[2]As in the linear case, nonlinear simultaneous equations must meet identification conditions. The complexities of this topic may be explored at an introductory level in Goldfeld and Quandt (1972, chap. 8). Recent advances include Brown (1983).

effect only when the value of x_1 is large and positive; the effects of y_2 and x_1 interact. Now the variable x_1y_2 is endogenous, since it includes y_2. That is, y_2 is correlated with the disturbance term, so that x_1y_2 will also be correlated. Hence this variable must somehow be purged.

The purging method requires knowledge of the *reduced form* of the variable y_2. In a simultaneous-equation system, there are two ways in which an exogenous variable can influence an endogenous variable—by direct causal effect or through influence on another endogenous variable that in turn affects the first one. The sum of all such impacts is the *total effect* of an exogenous variable; an equation that relates each endogenous variable to all the total effects acting upon it is the reduced-form equation for that variable. Every simultaneous-equation system implicitly defines a reduced form; that is, it implicitly specifies each endogenous variable as a function of the exogenous factors alone. (The reduced form is found simply by solving for the endogenous variables as functions of the exogenous ones.) The forecast values from this equation are needed for the estimation method.

When all equations are linear, consistent reduced-form forecasts can be obtained simply by regressing each endogenous variable on all the exogenous variables. This is the first stage or purging step of ordinary two-stage least squares. To achieve the statistical power of 2SLS in the interactive case mentioned above, the first step is to construct a consistent estimate of the reduced form forecasts of x_1y_2. In general, this is done by solving the complete system of equations for x_1y_2 in terms of x's alone. The following example is a simple case of this procedure.

Suppose that the complete model looks like this:

$$y_1 = a_1 + b_{11}x_1 + c_2x_1y_2 + u_1 \tag{1}$$

$$y_2 = a_2 + b_{12}x_1 + b_{22}x_2 + u_2 \tag{2}$$

where the y's are endogenous, the x's are exogenous, the u's are disturbances, and the usual conditions hold. The reduced form for y_2 alone is just Equation (2): it gives y_2 as a function of exogenous factors only, plus the disturbance term u_2. If this expression is substituted for y_2 in the term x_1y_2, the result is the reduced form:

$$x_1y_2 = a_2x_1 + b_{12}x_1^2 + b_{22}x_1x_2 + v_1 \tag{3}$$

where $v_1 = x_1u_2$ is a disturbance term with mean zero. A consistent estimate of the reduced form forecasts of x_1y_2 can thus be obtained simply by estimating Equation (3) as a regression equation and computing the forecast values.

Other kinds of interactive terms may be handled similarly. Suppose, for example, that the term y_2y_3 is added to Equation (1) as an explanatory variable. Suppose further that the model's equation for y_3 is:

$$y_3 = a_3 + b_{33}x_3 + u_3 \qquad (4)$$

Then the reduced form for y_2y_3 is the product of Equations (2) and (4), an equation of the form

$$y_2y_3 = h + k_1x_1 + k_2x_2 + k_3x_3 + k_4x_1x_3 + k_5x_2x_3 + v_2 \qquad (5)$$

where again h and the k's are fixed coefficients and v_2 is a disturbance term with mean zero. (Notice that the product of the disturbances u_2 and u_3 has nonzero mean in general, so that if v_2 is to have mean zero, $\text{cov}(u_2, u_3)$ must be grouped with the intercept in Equation (5).) Again a consistent estimate of the mean of the reduced form is available from the regression forecasts in (5).

Note that the dependent variable in (5) is the nonlinear term y_2y_3. A consistent estimate *cannot* be constructed by purging y_2 and y_3 separately and multiplying the purged values together. In general, purging the endogenous variable separately fails to estimate the reduced form of a nonlinear function of endogenous variables. When nonlinearities occur, endogenous terms must be purged as a single variable, not piecemeal (Kelejian, 1971).

Suppose, then, that in a nonlinear structural equation a consistent estimate of the reduced form forecast is available for every explanatory endogenous variable. Any consistent estimate will do; it need not be a regression forecast. For instance, Chapter 3 showed how to obtain the reduced form for a dichotomous endogenous variable generated by a probit equation. Then define the list of purging variables to be all the reduced-form estimates for nonlinear explanatory variables in the equation, plus the exogenous variables included in the equation.[3] (In Equation (1), for example, the list of purging

---

[3]It is assumed that the final list of purging variables is not collinear. That is, none can be expressed as a perfect linear function of the others.

variables would be the reduced-form forecasts for x_1y_2 plus x_1 and the intercept.)

With this list of purging variables, one can proceed just as in 2SLS. Purge each of the endogenous causal variables using these variables as regressors (for example, regress the quantity x_1y_2 on the purging variables). Then insert the forecasts into the original equation, apply regression, and correct the standard errors just as before. The result is estimates efficient within the class of single-equation instrumental variable estimators.[4] That is, in return for adding one extra step (the construction of the initial reduced-form estimates), the attractions of 2SLS can be extended to the nonlinear case. (The proof is given in the appendix to Chapter 3.)

In summary, then, the method of efficiently estimating an equation with nonlinear endogenous terms is as follows:

A. Construct a consistent estimate of the reduced form forecast of every linear and nonlinear endogenous explanatory variable. Define the list of purging variables to be these estimates plus the exogenous variables included in the equation.

B. Using this list of purging variables, carry out two-stage least squares.

NONRANDOMIZED EXPERIMENTS WITH ERRORS IN VARIABLES

Important social variables often cannot be measured exactly. In educational research, for example, researchers frequently control for scores on standardized tests; in survey research, beliefs or attitudes may be measured by the response to a single question. Variables of this kind contain measurement errors, sometimes rather large ones.

Both the econometric and psychometric literatures have demonstrated that if an independent variable in a regression equation is observed with error, ordinary least squares will not produce consistent

---

[4]In this procedure, the initial consistent estimates of the reduced forms can be replaced by the forecasts from regressing the endogenous variables on low-order polynomials in the exogenous variables. The 2SLS calculations remain consistent and will produce the correct standard errors, though with a loss of efficiency that can be quite large in certain cases. This technique is particularly helpful when the reduced form is unknown or hard to estimate. See the discussion of Kelejian's nonlinear two-stage least squares estimator in Chapter 3.

estimates of the coefficients (Sargan, 1958; Madansky, 1959; Lord, 1960; Werts and Linn, 1970, 1971; Kenney, 1975). Biases become particularly large when the measurement error is correlated with the dependent variable. The best-known instance occurs when an initial test score is used as an independent variable to predict a dependent variable that is a "change score"—that is, the difference between the first test and a later one. In this case, the errors in the first test enter both the independent and the dependent variables, and substantial biases can occur in estimating the effect of the first test.

In the ordinary regression case, measurement errors in dependent variables do not affect the unbiasedness or consistency of the estimates (if the errors have a mean of zero and are uncorrelated with the independent variables). Under the same conditions, measurement errors in endogenous variables in simultaneous equations cause no difficulties. These facts suggest that if a variable in a regression or in a structural equation is measured with error, it should simply be treated as endogenous whether or not it would be so regarded in the absence of error. That is, if erroneously measured variables occur, the equation should be estimated by two-stage least squares. Any variable with error should be purged by regressing it on all the exogenous variables in the system. Next the purged values should replace the original variable and ordinary regression should be applied, just as in any 2SLS procedure. The result is indeed consistent estimates of the coefficients. If the residual variance is corrected in the manner explained in Chapter 3, the standard errors computed by the usual regression calculations will also be correct.

In short, measurement error raises no new theoretical difficulties. Two-stage least squares copes with it in the same way that it deals with reciprocal causation. Even if the measurement error in the independent variables being purged is correlated with the dependent variable (as in the study of change scores), consistent estimates will result.[5]

---

[5]Again the equation being estimated must be identified. The variables with error are treated as endogenous; only variables not explained by the system that are also free from error count as exogenous. With that proviso, the same rules apply as before. In the ordinary regression case with one independent variable measured with error, for instance, there must be available at least one additional exogenous variable not included in the equation that can be used to predict the erroneously measured independent variable.

CENSORED SAMPLES WITH COMPLEX SELECTION SCHEMES

Chapter 5 discussed the estimation of regression relationships when all the observations obey the same selection equation. If selection follows probit or linear probability assumptions, then suitable estimates can be obtained simply by adding one additional variable to the regression equation under study, as explained in the last chapter.

In many applied problems, however, observations enter a censored sample in more than one way. For instance, a medical experiment might have data only from an experimental group composed partly of volunteers and partly of patients referred by physicians. If these two subgroups enter the sample for different reasons, as seems likely, then different selection equations will describe their entry. Hence not all observations follow the same selection rule, and the methods of Chapter 5 will be inappropriate.

A simple extension of the basic procedures, however, will cope with multiple entry paths in censored samples. Assume that every observation has two opportunities to enter the sample. (For example, patients have the chance to volunteer and the possibility of being referred.) Thus some of the observations enter the censored sample by one selection equation and the remainder by another, but every observation is subject to both equations. Assume further that the disturbances from these two equations are uncorrelated, so that if one controls for the independent variables, entering the sample in one way does not make an observation more or less likely to enter by the other route. (For example, unmeasured factors that create volunteers do not add or subtract from their chances to be recommended later by a doctor.) Thus some observations may qualify for sample entry in both ways. (Patients may volunteer *and* be referred by a physician.)[6]

As in Chapter 5, successful statistical use of a censored sample requires that something be known of how it was constituted. Suppose, then, that a sample of the population of interest is available,

---

[6]The assumption of uncorrelated disturbances is often approximately true even if entering the program in one way precludes entering it in another. If both groups of entrants are small fractions of the eligible population and the selection equations forecast that no one has a good chance of being chosen in either way, then assuming the disturbances uncorrelated implies that very few entrants would be selected on both criteria. This implication gives a reasonable fit to the sample, in which no one is so chosen.

consisting of the censored sample (for which outcomes are known) plus other observations not selected for the program (whose outcomes are of course unknown). In the medical example, the censored sample would consist of those patients who volunteered or were recommended for the experimental treatment; the others would be people with the same disease outside the program. Separate regressions (or probit equations) are then estimated for the probability of being chosen for the sample in either of the two ways. (One equation would forecast the chance that a subject will volunteer, for example, the other that the person will be referred by a doctor.) Then when regression is applied to the censored sample, instead of adding one additional variable from the single selection equation as in Chapter 5, two variables are added, one from each of the selection equations. If selection occurs according to the assumptions of the linear probability model, the residuals from both of the selection equations would be added to the outcome equation as additional variables. One residual might be positive and one negative, for instance, meaning that the subject was accepted on one criterion but not on the other (say, a nonreferred volunteer). Ordinary regression is then applied; under natural extensions of the assumptions in Chapter 5, consistent estimates result. A similar procedure follows when selection obeys probit assumptions.[7]

In other instances, distinct populations are eligible for different selection mechanisms. In the medical example, members of the local community might be accepted only if they volunteer whereas nonresidents would be eligible only if referred. Suppose that, for the purposes of the experiment, these two populations are medically equivalent apart from the accident of geography, so that both can be described by the same outcome equation. In that case, one proceeds as above except that selection equations are estimated separately for the relevant groups—one equation for city residents, for example, and another for out-of-towners. Once again, two variables will be added to the outcome equation, but the variable from the inappropriate

---

[7]When a subject is accepted into the sample in one of the probit selection equations, the variable to be added to the outcome equation is λ, as explained in Chapter 5. When the subject fails to be accepted, the variable to be added is $-\varphi/\Phi$, where these quantities are defined as in Chapter 5.

selection equation would be set to zero. Ordinary regression again will be consistent.

These extensions of the elementary censoring model by no means exhaust the possibilities, and it is useful to have a sense of their limitations. More complex selection procedures are easily imagined and do occur in practical situations. In the study of pretrial release, for example, arrestees may get out of jail prior to trial if they are released on recognizance. Failing that, bail is set and release may be obtained by posting the required sum. Thus accused criminals may enter the released group in one of two ways—recognizance or bail. The disturbances from the two are probably correlated: the same ties to the community and economic stability that persuade a judge to release on recognizance will also help raise bail money. Moreover, the two processes are sequential and coupled: no one attempts to raise bail who has been released on recognizance. Thus if a criminologist wishes to use the sample of released defendants to draw conclusions about the full population of defendants, none of the methods considered so far is adequate.

Another variation on the censored sample is *retrospective sampling*, in which sampling is stratified by outcomes. A medical experiment with many failures and few successes, for example, may oversample the successes in an effort to learn how to increase their frequency. Thus the chance of entering the sample depends on the outcome, not just on exogenous factors as in the examples of Chapter 5. (See also Manski and Lerman, 1976.)

Yet another complexity occurs when retrospective sampling is combined with simultaneous causation within the selected group. An intriguing study by Blechman and Kaplan (1978) asked this question: Faced with a foreign policy crisis, what kind of threatening military moves (if any) should American officials make to maximize their chance of success? Presumably the probability of success influences the threats chosen and the threats influence the chance for success, so that reciprocal causation is at work. Moreover, the authors have data only for instances in which some military activity, but no actual hostilities, was employed—meaning that outcomes such as actual fighting, purely verbal threats, and simply doing nothing were excluded. Thus the sample is censored. The complete model constitutes a simultaneous-equation system that is observed only for a nonrandom fraction of the population of interest.[8]

The pretrial release, retrospective sampling, and foreign policy examples cannot be adequately handled with the methods set out thus far. Estimation procedures can be devised for them, but these methods involve techniques beyond the level of this book.[9]

TIME SERIES PROBLEMS

One large class of research neglected thus far deals with observations taken over a period of time. A researcher might examine European drunk driving arrests and auto accidents for several years to assess the effect of a tougher law against drinking and driving (Ross, 1975), for example, or analyze monthly property crimes in Detroit to decide whether drug treatment programs reduce addicts' incentive to steal (Levine, Stoloff, and Spruill, 1976). The great advantage of such studies is that past behavior of the subjects usually serves as a good control for their current activities: a jurisdiction that was full of drunk drivers or heroin addicts last month is very likely to possess them in abundance this month as well. If a data set is composed of repeated observations on essentially the same subjects, many of the difficulties of modeling are avoided. The long lists of control variables needed to make geographic areas comparable in cross-sectional studies are obviated when the same jurisdiction generates all the data.

In the simplest case, time series studies can be analyzed like any other quasi-experiment. Suppose that unobserved factors influencing outcomes do not persist from one time to another—that is, in a regression equation describing outcomes, the disturbances are uncorrelated. In time series parlance, no "serial correlation" occurs.

---

[8]Blechman and Kaplan (1978, pp. 18–19) make the ingenuous claim that by restricting themselves to simple analyses (contingency tables) they minimize the statistical biases in their sample. Precisely the reverse is true. One important reason why case studies like this are so unsuccessful is that the ad hoc methods used to draw the sample virtually ensure that any simple analysis (statistical or verbal) will quickly lose itself in the artificial "patterns" created by the sampling procedure.

[9]Still other kinds of censored samples are produced by mechanisms unlike those considered in this book. For instance, the New Jersey Income Maintenance Experiment selected subjects on the basis of income, a continuous variable. The resulting selection equation is of the Tobit form, rather than probit or linear probability. For discussion of such models and their estimation, see Maddala (1983).

The techniques of previous chapters then apply directly. In the drunk driving example, the policy change to be studied is a new law. If the measure of success is the number of drunk driving arrests, the causal variables influencing this measure might include drunk driving arrests the previous month (a lagged dependent variable), a variable for the treatment (1 if the new law was in effect, 0 otherwise), plus whatever control variables prove appropriate (month of the year, total miles driven, and so on). This constitutes the outcome equation.

As in cross-sectional studies, the statistical nature of the treatment variable in this equation requires close inspection. If the new law was put into effect for reasons having nothing to do with the number of drunk driving arrests (the dependent variable), the outcome equation can be estimated as an ordinary regression problem. More precisely, if the starting date of the law was influenced neither by the number of arrests nor by unmeasured factors influencing arrests, then the presence of the law is "predetermined" and can be treated as if it were exogenous. The sample can be dealt with as though the law had been imposed in a random month, and ordinary regression will be unbiased and consistent.

More commonly, however, policy changes in time series studies are endogenous—and for precisely the same reasons as in cross-sectional research. Laws are changed when the need arises. In Campbell and Ross's (1968) classic study of Connecticut's crackdown on speeding, enforcement of the speed limits was tightened precisely because an extraordinary number of highway deaths had occurred. If the existence of the law is due only to the death rate in *previous* time periods, regression will be satisfactory. But if reciprocal effects occur rapidly or the intervals at which the data are available are fairly wide, the impact of treatment and outcome on each other may be essentially simultaneous. In either case, two-stage least squares and related methods are required if consistent estimates of the treatment effect are to be obtained. In the absence of serial correlation, the time series nature of the data causes no additional problems. The methods of Chapter 3 suffice for consistent estimates.

More serious difficulties appear when, as frequently happens, serial correlation affects outcomes. Suppose first that outcomes are continuous (not just "success" or "failure") and that the experiment is essentially randomized—that is, assignments to experimental groups are neither caused by outcomes nor influenced by the same unobserved factors that affect outcomes. Assume further that previous outcomes do not affect current outcomes. Then a quasi-experiment of this type is an ordinary regression problem in every respect except the serial correlation: unobserved factors that influence outcomes at one time period survive to influence them at subsequent time periods as well.

In the most elementary case, unobserved variables persist through a single time period—first-order serial correlation. This means simply that the current disturbance is not directly influenced by any disturbance other than the previous one. Put the other way, unobserved factors have a direct impact just one time period ahead—often an adequate approximation in practice.

Ordinary regression gives a consistent estimate when disturbances are subject to first-order correlation, but the estimates are not efficient and the computed standard errors are incorrect. A variety of correction procedures are given in elementary econometrics texts; the following approach is perhaps the simplest.

To begin, suppose that an estimate r has somehow been obtained for the correlation between each disturbance and its predecessor. (Methods of doing so are discussed below.) This is a simple Pearsonian correlation coefficient. Then define the *differenced value* for any variable as its current value minus the product of r with the previous value. If y_t is the current value of the dependent variable and y_{t-1} is the previous value, for example, the differenced value is

$$y_t^* = y_t - ry_{t-1}$$

Differenced values are defined for independent variables in the same way. In particular, the intercept term, which equals 1 for all observations, has a differenced value of $1 - r(1) = 1 - r$. The estimation then proceeds in this way:

A. Apply ordinary regression to the original equation.
B. Compute the correlation between residuals and their predecessors in step A. This gives the estimate r. For every variable in the regression, including the intercept, compute the differenced values.
C. Apply regression again, regressing the differenced dependent

variable on the differenced independent variables, including the differenced intercept. Suppress the usual intercept.[10]

Under conventional regression assumptions, the result of this procedure is asymptotically efficient estimates whose standard errors are appropriate.

In practice, the assumptions behind this simple procedure frequently fail. Reciprocal causation or disturbances correlated across the selection and outcome equations enter time series studies just as they do cross-sectional quasi-experiments. In a study of traffic law enforcement, for example, common sense suggests both that crackdowns occur in response to unsatisfactory current safety levels and that whatever caused the highways to be unsafe in the last period is likely to continue to make travel hazardous in the near future. Then both simultaneity and serial correlation are at work.

Perhaps the most troublesome feature of serial correlation in simultaneous equations is its effect on lagged values of the endogenous variables. Without serial correlation, values of the outcome or treatment variables from previous time periods are exogenous and can be employed in a regression as independent variables without compromising the statistical conclusions. When disturbances are correlated, however, this rule no longer applies. Lagged endogenous variables must be treated as endogenous, since they are correlated with the disturbances in the equation.

More elaborate methods are required to cope with these complexities. Again only the first-order serial correlation case is discussed here and just one simple estimator is explained. The methods are set out only for the case in which no endogenous variable in the equation is lagged more than one time period; that is, all the causal variables are observed either at the current time period or at the one preceding. No variable has direct effects more than one period ahead. (This restriction can be relaxed; see Kelejian and Oates, 1974, pp. 261–64.)

To apply this estimator, first define the augmented list of exogenous variables to be all the exogenous variables at the current time period

---

[10]Alternatively, the differenced intercept term can be omitted and the usual intercept included so long as the coefficient for the latter and its standard error are divided by $1 - r$ at the end.

plus all those from the preceding time period. Then the procedure is as follows:

A. Using the augmented list of exogenous variables, apply two-stage least squares. Both current and lagged endogenous variables must be purged.[11] The result will be consistent (but inefficient) coefficient estimates, and the standard errors will be wrong.
B. Reinserting the true values of the endogenous variables in place of their purged values, compute the residuals in the second stage of step A. Estimate r by correlating the residuals with the same residuals lagged one period.
C. Compute the differenced values for all variables, including the intercept. (See the preceding discussion of the ordinary regression case for the construction of these variables.) Note that if a variable was initially lagged one period, its differenced value is the original lagged variable less the product of r with the same variable lagged *two* periods—for example, $y^*_{t-1} = y_{t-1} - ry_{t-2}$.
D. Apply 2SLS, using the differenced values and correcting the standard errors at the end as usual. Under the standard assumptions, the result is asymptotically more efficient estimates whose standard errors are correct.

When time series data follow more complex laws—second or higher-order serial correlation, perhaps combined with endogenous variables lagged several periods—alternative techniques become necessary. If no simultaneity is present, models of this type may be estimated by Box–Jenkins methods (Box and Jenkins, 1970; elementary introductions include Cook and Campbell, 1979, and Glass, Willson, and Gottman, 1975). Extending these methods to deal with reciprocal causation requires techniques beyond the scope of this book.

Finally, although censored samples occur only rarely in time series contexts, practical examples do exist of processes occurring over time

---

[11]It is assumed that with the lagged endogenous variables counted as additional endogenous variables, the equation is identified.

and observed only at certain time points. For instance, a study of a community's history of conflict that is based on newspaper reports might have data only for disputes that reached a certain level of newsworthiness. The procedures of this and the previous chapter no doubt can be extended to this situation, but apparently no such estimators have yet been proposed.

DATA ANALYSIS

Throughout this book, primary emphasis has been placed on the statistical aspects of the analysis of quasi-experiments. Computational routines have been set out which, under certain conditions, yield statistically attractive estimates. In every instance, these procedures generalize simpler methods such as cross-tabulation or regression—in the sense that the more sophisticated techniques give trustworthy estimates whenever the elementary methods do—and they also cope with common features of applied research that defeat cross-tabulation and regression. At an abstract level, therefore, this book constitutes one example of how econometric theory provides "solutions" to the conundrum of nonrandomization.

Theoretical results of this sort are essential. Without theory, data analysis quickly uses up the simplest techniques and finds itself overwhelmed by choices of what to do next—no decision is evidently better than another. With theory, on the other hand, the researcher can transcend the limits of intuition.

Yet mathematics alone is clearly insufficient also. Applied statistics relies on statistical theory only as a kind of moral guidance. No realistic data set meets all the assumptions an econometrician must make. Empirical work, to a greater or lesser degree, falls short of the commandments set out in textbooks. The good researcher, then, does not delude himself that he is without fault. Instead, knowing that the fall from grace is inevitable, he aims to avoid the truly blameworthy and to ignore mere peccadillos. Discovering which is which defines what data analysis is about.

Almost by definition, good data analysis cannot be set out as a system of rules. Practical judgments apply to unique cases. Yet general guidelines can be given; and in that spirit, suggestions are offered here on recurrent data-analytic topics in nonrandomized experiments and censored samples. Exhaustiveness is neither desirable nor feasible.

Instead, a variety of topics are touched on to illustrate the range of techniques available. The treatment is brief; these and other data-analytic methods are best learned by trying them out.

THE LINEAR PROBABILITY MODEL VS. PROBIT

Dichotomous endogenous variables demand special econometric techniques. (McFadden, 1974, is a good overview.) Two of the most popular of these approaches are the linear probability model and probit analysis. Both can be extended to cover statistical complexities beyond those for which they were developed, and both figured prominently when statistical techniques for quasi-experiments with nonrandomized control groups and with censored samples were discussed earlier in the book.

As was explained in Chapters 3 and 5, the choice between these methods turns on considerations of elegance versus simplicity. The probit model more easily handles "floor and ceiling effects"—in essence reducing the impact of variables when they threaten to lead to probabilities greater than 1 or less than 0. This feature, combined with the somewhat more plausible theoretical structure of probit analysis, makes it a more sophisticated model conceptually.

In practice, however, probit analysis is computationally slow and therefore expensive, especially in structural equation models where several rounds of probit may be needed to estimate a single equation. Moreover, the standard errors may have to be estimated separately. In the exploratory phases of empirical work, when equations may be estimated ten, twenty, or even fifty times, probit analysis frequently prices itself out of reach. In those circumstances, the linear probability model becomes more attractive. Aesthetics are sacrificed in return for computational efficiency.

Whether a researcher should turn to the linear probability model depends, of course, on the characteristics of the data (Warner, 1976). At one extreme, the loss of fit may be negligible or even zero. Consider, for example, the case of a dichotomous dependent variable and two independent variables, each of which has just three categories. Observations may fall into any of three categories on both of the two latter variables. In effect, then, the problem is to fit a 3×3 table. The entry in each cell of the table is the proportion of observations that are scored 1 on the dependent variable.

A "saturated" probit model for this problem would employ eight independent variables plus an intercept term, one variable for each of the nine categories of the table. For example, each of the eight variables other than the intercept could be scored as dummy variables corresponding to cell categories—if the observation fell in that cell, 0 otherwise. The intercept would represent the ninth cell. A linear probability model could be defined in exactly the same way. In fact, the forecasts from the probit and linear probability models would be *identical* in this case. The cheaper and simpler linear probability model will show no probability forecasts outside the appropriate range and will incorporate whatever floor and ceiling effects are present in the observations. It will "explain" the data every bit as well as the probit model.

This example shows that the theoretical elegance of the probit model need not imply superior fit in practice. Its conceptual appeal is meaningless in certain situations. In fact, forecasts equal or arbitrarily close to those of the probit model can always be achieved with the linear probability method by adding cross-product and polynomial terms in the independent variables to the regression equation. When most of the independent variables are discrete, as in the preceding example, or if the true probabilities are largely confined to either the lower, the upper, or the central part of the 0 to 1 range, then just a few polynomial and interaction terms in the key variables, or even none at all, may give forecasts essentially identical to those of the probit model. A more cluttered equation results whose interpretation may require a bit more care, but except for the possible loss of a few degrees of freedom, it will be a statistical synonym for the corresponding probit equation if the latter is the correct model (and may well improve on it if it is not).

When, then, is probit methodology likely to improve fit? The linear probability model is at its worst when true probabilities cover most of the possible range and when simultaneously many of the independent variables are continuous. At this extreme, the number of polynomial and interactive terms necessary for verisimilitude in the linear probability model becomes unwieldy, especially if the number of original independent variables is already quite large. Yet without these terms, forecasts well outside the meaningful range will occur (see, for example, Aldrich and Cnudde, 1975). In these circumstances, probit analysis ordinarily justifies its cost. Despite its complexity, it becomes simpler to use than the linear probability model.

Most practical situations fall between the two examples given above, of course, making the choice between probit and the linear probability model one of relative costs. With a bit of care to ensure that additional variables are added to the linear probability model to eliminate meaningless forecasts, in most ordinary single or multiple-equation contexts the fit of the two models can be made virtually equivalent. Whether this extra effort at specification and interpretation outweighs the computational costs of probit analysis then becomes largely a matter of personal preference.

SPECIFICATION SEARCHES

In most quasi-experiments, analysts do not enter upon their studies with fixed, well-understood variable lists. They may not know whether some variables have large or small effects, and the functional form of those impacts may also be obscure. (The best theoretical treatment is Leamer, 1978.) A period of searching for the appropriate causal structure is part of most empirical social studies. There are several good practical texts on this subject (see Daniel and Wood, 1971; Mosteller and Tukey, 1977; Tukey, 1977), and the methods they set out for regression models apply with obvious modifications to structural equation estimation, including the techniques of this book. Rather than listing these obvious analogs, two specification issues unique to multi-equation contexts will be discussed instead—the choice of excluded variables in structural equations and tests of the distributional form of the disturbances in censored samples.

In a quasi-experiment, the absence of randomization must be compensated for by other restrictions on the data. Usually the researcher must have in hand at least one variable with some influence on selection but none on outcomes. Often no single variable will be *known* to have this property; instead a number of candidates will be in nomination, only some of which will withstand scrutiny. The analyst must choose among them.

Essentially the rules for choosing a good specification in a two-stage least squares or censored-sample problem are the same as those in ordinary regression. So long as the appropriate exclusion restrictions are maintained, a variety of specifications can be tried and compared for statistical power and theoretical or a priori plausibility. If important variables have similar effects across the reasonable formulations, then confidence in the results is strengthened. In particular, if an equation is overidentified (there are more excluded variables than required), the excluded variables can be tried out one or a few at a time to assess their credibility. This is not a formal test of the hypothesis that the equation is identified, but it often provides much informal information and assurance.

Any of the stepwise regression methods, if not followed slavishly, can be of service in the search for excluded factors. Variables that enter an equation with a large coefficient and a substantial increase in explanatory power are poor candidates for exclusion. Of course, all such procedures presume that the search has been limited. An initial list of reasonable excluded factors must be separated from variables known to enter the equation. It would be foolish, for example, to load every variable into a stepwise algorithm and let the algorithm decide which to include or exclude. Some prior knowledge is essential.

The second specification issue arises in censored samples. Here the usual specification issues are supplemented by those stemming from the covariance between the disturbances in the outcome and selection equations. Chapter 5 showed that probit assumptions lead to the addition of a variable λ to the equation, whereas the linear probability approach implies that the residual from the selection equation, $\hat{u}_1$, should be added. When most observations in the sample have entry probabilities of 20 percent or more, the distinction between the probit and linear probability models makes little difference. However, samples of data points that have relatively little chance of appearing require more care.

Individuals who are unlikely to enter a sample but arrive anyway are highly distinctive by definition. Some unobservable factors have caused them to behave differently. In models for censored data, the mean effect of these unknown variables is estimated from the tail area of a probability distribution. The more unlikely the entry of the observation, the more extreme the tail area considered. Tail areas, however, are precisely the part of a curve in which simplifying distributional assumptions such as normality or uniformity are most likely to mislead. Unlike the usual structural equation estimators such as those discussed in Chapter 3, which can be justified with postulates that do not specify the exact functional form of the distribution of the disturbances, censored-sample methods depend quite strongly on knowledge of that distribution. Thus when most observations in a censored sample have low probability of entry, the resulting estimates depend quite strongly on some of the most arbitrary features of the model.

At least partial remedies are at hand, however. In any statistical model whose true functional form is unknown, data analysts routinely try out a variety of them to select those with the best fit. Variables may be transformed, for example, and quadratic and cubic terms may be added to a regression equation. The same procedures protect a study of a censored sample against error in its distributional assumptions. Adding, say, both λ and λ^2 to an outcome equation provides considerable protection against specification errors, for example, as will adding both $\hat{u}_1$ and $1/\hat{u}_1$. A dummy variable for observations with very low probability of entering the sample can also be added, or a series of them, one for each of several probability levels. If the additional variables prove useless, either because they acquire zero coefficients or because they leave the other coefficients unchanged, they can be omitted. Informal methods of this kind provide a certain robustness to censored-sample methods and protect against gross errors deriving from overreliance on distributional assumptions.

WHEN REGRESSION IS ENOUGH

Throughout this book the principal emphasis has been placed on the shortcomings of ordinary regression and on techniques for remedying them. Of course, regression and other familiar methods such as cross-tabulation may sometimes be perfectly satisfactory, even in quasi-experiments with nonrandomized assignment or with censored data. This is most clearly the case when the results of correcting for nonrandomized assignment or censoring indicate that the corrections are unnecessary.

In the models of Chapter 3, for example, if residuals between equations are very nearly uncorrelated when the quasi-experiment is analyzed by two-stage least squares, regression will perform satisfactorily. The differences between the treatment and control groups have been captured adequately by the existing exogenous variables. Similarly, if in a censored sample the coefficient on λ or $\hat{u}_1$ is small, then few unmeasured factors are common to both the selection and outcome equations. Again the independent variables in the outcome equation control for most of the factors that also influence selection; the data set resembles a stratified sample.

In all such cases, ordinary regression is only slightly inconsistent and its coefficients resemble those of elaborate methods within the limits of sampling error. Simple regression is preferable, simply because its stronger assumptions produce smaller standard errors for the coefficients. The methods of this book then serve only to verify these assumptions at a preliminary stage.

In other circumstances, a researcher may revert to regression under less happy circumstances. All the procedures in this book designed to remedy the limitations of regression depend on additional assumptions about the data, usually the claim that certain variables influential in selection make no difference in outcomes. In some instances good excluded factors simply do not exist. Symptoms may appear in the specification searches, and variables thought to be excluded may show important explanatory power when added to the outcome equation. Alternatively, the coefficient estimates may be quite unstable and the standard errors large—meaning that the excluded variables probably make too little difference in the selection equation to be good exclusions in explaining outcomes.

Similarly, if every observation in the censored-sample problem is estimated to have approximately the same probability of entering the sample, then any outcome equation with an intercept term will be estimable only with difficulty. That is, the residual $\hat{u}_1$ from the selection equation will be nearly constant and therefore close to collinear with the intercept. Because the factors in the selection equation explain so little variation, and hence not much is known about entry into the sample, correcting for censoring becomes difficult.

When structural equation methods give highly noisy estimates or collapse entirely, the simpler cross-tabulation and regression methods are the only alternative. The price is bias and inconsistency, but the direction of the error may be guessed at—and in any case an inconsistent estimator with small variance may be closer to the truth than a very uncertain consistent estimator. No attractive choice exists here, of course, and the selection of regression occurs by default.

Considerations like these bring to mind again the power of true experiments. Randomization is a powerful device, and its loss makes valid inference difficult—more substantive knowledge is needed and more information about concomitant variables. Without extra assumptions and data, quasi-experiments are suggestive at best, and more often useless or downright misleading, no matter what statistical computations are attempted. When possible, randomized experiments are always preferable.

The message of this book, however, is that even when randomization is impossible, researchers are often not without recourse. Many quasi-experiments present no insurmountable obstacles to rigorous inference. Under the right conditions, notably when certain measured variables are known to have a strong effect on assignment to treatments (or sample selection) but no effect on outcomes, in principle many nonrandomized studies can be analyzed statistically with the same confidence that attaches to pure experiments. And in practice, the data-analytic issues that arise are little different from those presented by classical experimentation.

Much remains to be done, of course, before quasi-experimental methodology attains full maturity. We lack statistical theory for the more complex quasi-experimental designs and informal understanding of the data-analytic issues posed by nonrandomized data. Both theory and intuition need further development. For nonrandomized designs are inescapable: many theoretically intriguing and humanly important topics can be studied in no other way. Extending our ability to learn from quasi-experiments, then, is fundamental to the progress of the social and medical sciences.

Bibliography

Achen, Christopher H. 1980. "Estimating Treatment Effects in Quasi-Experiments: The Case of Censored Data." Berkeley: Department of Political Science, University of California.

Aldrich, John, and Charles Cnudde. 1975. "Probing the Bounds of Conventional Wisdom." *American Journal of Political Science* 19: 571-608.

Amemiya, T. 1973. "Regression Analysis When the Dependent Variable Is Truncated Normal." *Econometrica* 41, 6 (November): 997–1017.

______. 1978. "The Estimation of a Simultaneous Equation Generalized Probit Model." *Econometrica* 46: 1193–205.

Angel, Arthur R., et al. 1971. "Preventive Detention: An Empirical Analysis." *Harvard Civil Rights–Civil Liberties Review* 6, 2 (March): 300–96.

Astin, Alexander W. 1977. *Four Critical Years*. San Francisco: Jossey-Bass.

Barnow, B. S. 1972. "Conditions for the Presence or Absence of a Bias in Treatment Effect: Some Statistical Models for Head Start Evaluation." Madison: Institute for Research on Poverty Discussion Paper 122-72, University of Wisconsin.

Bases, Nan C. 1972. *Preventive Detention in the District of Columbia: The First Ten Months.* Washington, D.C.: Georgetown Institute of Criminal Law and Procedure.

Basmann, R.L. 1957. "A Generalized Classical Method of Linear Estimation of Coefficients in a Structural Equation." *Econometrica* 25: 77–83.

Beeley, Arthur L. 1927. *The Bail System in Chicago*. Chicago: University of Chicago Press.

Berman, Paul. 1974. *Revolutionary Organization*. Lexington, Massachusetts: D.C. Heath.

Blechman, Barry, and Stephen S. Kaplan. 1978. *Force Without War: U.S. Armed Forces as a Political Instrument.* Washington, D.C.: Brookings.

Boruch, Robert F. 1976. "On Common Contentions About Randomized Field Experiments." In Gene V. Glass, *Evaluation on Studies Review Annual,* vol. 1. Beverly Hills: Sage Publications.

Bowers, William G. 1974. *Executions in America*. Lexington, Massachusetts: D. C. Heath.

Box, G. E. P., and G. M. Jenkins. 1970. *Time-Series Analysis: Forecasting and Control.*. San Francisco: Holden Day.

Bronfenbrenner, Urie. 1975. "'Is Early Intervention Effective?" In Elmer L. Struening and Marcia Guttentag, *Handbook of Evaluation Research*, vol. 2. Beverly Hills: Sage Publications.

Brown, B.W. 1983. "The Identification Problem in Systems Nonlinear in the Variables." *Econometrica* 51: 175–96.

Cain, G.G. 1975. "Regression and Selection Models to Improve Nonexperimental Comparisons." In C. A. Bennett and A. A. Lumsdaine, eds. *Evaluation and Experiment: Some Critical Issues in Assessing Social Programs.* New York: Academic Press.

Campbell, Donald T., and R.F. Boruch. 1975. "Making the Case for Randomized Assignment to Treatments by Considering the Alternatives." In C. A. Bennett and A. A. Lumsdaine, eds., *Evaluation and Experience: Some Critical Issues in Assessing Social Programs.* New York: Academic Press.

Campbell, Donald T., and Albert Erlebacher. 1975. "How Regression Artifacts in Quasi-Experimental Evaluations Can Mistakenly Make Compensatory Education Look Harmful." In Elmer L. Struening and Marcia Guttentag, *Handbook of Evaluation Research*, vol. 1. Beverly Hills: Sage Publications.

Campbell, Donald T., and H. Laurence Ross. 1968. "The Connecticut Crackdown on Speeding: Time-Series Analysis Data in Quasi-Experimental Analysis." *Law and Society Review* 3, 1: 33–53

Campbell, Donald T., and J. C. Stanley. 1963. "Experimental and Quasi-Experimental Designs for Research on Teaching." In N. L. Gage, ed., *Handbook of Research on Teaching*. Chicago: Rand McNally.

Cicirelli, V., et al. 1969. *The Impact of Head Start.* Athens, Ohio: Westinghouse Learning Corp., Ohio University.

Clarke, Stevens H., Jean L. Freeman, and Gary G. Koch. 1976. "Bail Risk: A Multivariate Analysis." *Journal of Legal Studies* 5, 2 (June): 341–85.

Cochran, W. G. 1968. "The Effectiveness of Adjustment by Subclassification in Removing Bias in Observation Studies." *Biometrics* 24, 1 (March): 295–313.

______. 1983. *Planning and Analysis of Observational Studies.* New York: Wiley.

Coleman, James S., et al. 1966. *Equality of Educational Opportunity.* Washington, D.C.: U.S. Government Printing Office.

Coleman, James S., Thomas Hoffer, and Sally Kilgore. 1982. *High School Achievement*. New York: Basic.

Connor, Ross F. 1978. "Selecting a Control Group: An Analysis of the Randomization Process in Twelve Social Reform Programs." In Thomas Cook, Marlyn L. Del Rosario, Karen M. Hennigan, Melvin M. Mark, and William M.K. Trochim, eds., *Evaluation Studies Review Annual* 3: 104–53. Beverly Hills: Sage.

Cook, Thomas D., and Donald T. Campbell. 1979. *Quasi-Experimentation*. Chicago: Rand McNally.

Daniel, Cuthbert, and Fred S. Wood. 1971. *Fitting Equations to Data.* New York: Wiley. 2nd ed., 1980.

Dawes, R. M. 1975. "Graduate Admissions Criteria and Future Success." *Science* 187: 721–23.

Dhrymes, Phoebus. 1970. *Econometrics.* New York: Harper and Row.

District of Columbia Bail Agency. 1977. "Report of the D.C. Bail Agency." Washington, D.C.: District of Columbia Bail Agency.

Dykstra, A. D. 1971. "Logic of Causal Analysis: From Experimental to Nonexperimental Designs." In Hubert M. Blalock, ed., *Causal Models in the Social Sciences,* 273–94. Chicago: Aldine-Atherton.

Eklund, Gunnar. 1960. *Studies of Selection Bias in Applied Statistics.* Stockholm: Almqvist and Wiksells.

Executive Board of the National Conference on Bail and Criminal Justice. 1965. *Proceedings of the Institute on the Operation of Pretrial Release Projects.* New York: U.S. Department of Justice.

Feinberg, Lawrence. 1978. "Desegregation Discounted." *Washington Post*, September 18, p. A5.

Finney, D. J. 1947. *Probit Analysis*. Cambridge: Cambridge University Press. 3rd ed., 1971.

Fisher, Franklin. 1966. *The Identification Problem in Econometrics.* New York: McGraw Hill.

Foote, Caleb. 1954. "Compelling Appearance in Court: Administration of Bail in Philadelphia." *University of Pennsylvania Law Review* 102: 1031–79.

Freed, Daniel J., and Patricia M. Wald. 1964. *Bail in the United States: 1964*. Report to the National Conference on Bail and Criminal Justice, Washington, D.C., May 27–29, 1964. Washington, D.C.: U.S. Department of Justice.

Friedland, Martin L. 1965. *Detention Before Trial*. Toronto: University of Toronto Press.

Gilbert, John P., Richard J. Light, and Frederick Mosteller. 1975. "Assessing Social Innovations: An Empirical Basis for Policy." In Carl A. Bennett and Arthur A. Lumsdaine, eds., *Evaluation and Experiment*, 39–193. New York: Academic Press.

Gilbert, John P., B. McPeek, and Frederick Mosteller. 1977. "Statistics and Ethics in Surgery and Anaesthetics." *Science* 198: 684–89.

Glass, G. V., V. L. Willson, and J. M. Gottman. 1975. *Design and Analysis of Time-Series Experiments*. Boulder, Colorado: Colorado Associated University Press.

Goldberger, Arthur S. 1961. "Stepwise Least Squares: Residual Analysis and Specification Error." *Journal of the American Statistical Association* 56, 296 (December): 998–1000.

———. 1964. *Econometric Theory*. New York: Wiley.

———. 1972. "Selection Bias in Evaluating Treatment Effects: Some Formal Illustrations." Institute for Research on Poverty Discussion Paper 123-72. Madison: University of Wisconsin.

Goldberger, Arthur S., and D. B. Jochems. 1961. "Note on Stepwise Least Squares." *Journal of the American Statistical Association* 56, 293 (March): 105–10.

Goldfeld, S. M., and R. E. Quandt. 1972. *Nonlinear Methods in Econometrics*. Amsterdam: North Holland.

Haavelmo, Trgve. 1947. "Methods of Measuring the Marginal Propensity to Consume". *Journal of the American Statistical Association* 42 (March): 105–22.

Haberman, Shelby. 1974. *The Analysis of Frequency Data*. Chicago: University of Chicago Press.

Hansen, Lars Peter. 1982. "Large Sample Properties of Generalized Method of Moments Estimators." *Econometrica* 50: 1029–54.

Hanushek, Eric A., and John E. Jackson. 1977. *Statistical Methods for Social Scientists*. New York: Academic Press.

Hanushek, Eric A., and John F. Kain. 1972. "On the Value of *On Equality of Educational Opportunity* as a Guide to Public Policy." In Frederick Mosteller and Daniel P. Moynihan, *On Equality of Educational Opportunity*. New York: Vintage Press.

Hausman, J. A., and D. A. Wise. 1976. "The Evaluation of Results from Truncated Samples: The New Jersey Negative Income Tax Experiment." *Annals of Economic and Social Measurement* 5: 421–45.

Heckman, James. 1974. "Shadow Prices, Market Wages, and Labor Supply." *Econometrica* 42, 4 (July): 679–94.

———. 1976. "The Common Structure of Statistical Models of Truncation, Sample Selection, and Limited Dependent Variables and a Simple Estimator for Such Models." *Annals of Economic and Social Measurement* 5, 4: 475–92.

———. 1978. "Dummy Endogenous Variables in a Simultaneous Equation System." *Econometrica* 46: 931–59.

———. 1979. "Sample Selection Bias as a Specification Error." *Econometrica* 47: 153–61.

Hillinger, Charles. 1979. " 'Closet' Southpaws Found More Prone to Disease." *Los*

Angeles Times, July 5, part II, p. 4.

Holt, Norman. 1976. "Rational Risk Taking: Some Alternatives to Traditional Correctional Programs." In Gene V. Glass, *Evaluation Studies Review Annual*, vol. 1. Beverly Hills: Sage Publications.

Hsiao, Cheng. 1983. "Identification." In Zvi Griliches and Michael D. Intriligator, eds., *Handbook of Econometrics* 1: 223–83. Amsterdam: North Holland.

Jackman, Robert W. 1976. "Politicians in Uniform." *American Political Science Review* 70, 4 (December): 1078–97.

Jorgenson, D. W., and J. Laffont. 1974. "Efficient Estimation of Nonlinear Simultaneous Equations with Additive Disturbances." *Annals of Economic and Social Measurement* 3: 615–40.

Kelejian, H. H. 1971. "Two-Stage Least Squares and Econometric Systems Linear in Parameters but Nonlinear in the Endogenous Variables." *Journal of the American Statistical Association* 66: 373–74.

Kelejian, Harry H., and Wallace E. Oates. 1974. *Introduction to Econometrics*. New York: Harper and Row. 2nd ed., 1981.

Kendall, Maurice G., and Alan Stuart. 1969. *The Advanced Theory of Statistics*. Vol. 1, 3rd ed. New York: Hafner.

Kenney, David A. 1975. "A Quasi-Experimental Approach to Assessing Treatment Effects in the Nonequivalent Control Group Design." *Psychological Bulletin* 82, 3: 345–62.

King, Wayne. 1978. "Studies Show South Still Tougher on Black Criminals." *Minneapolis Tribune*, November 20, p. 1A. (New York Times News Service)

Kirby, Michael P. 1977. "The Effectiveness of the Point Scale." Washington, D. C.: Petrial Services Resource Center.

Kish, Leslie. 1965. *Survey Sampling*. New York: Wiley.

Landes, William A. 1974a. "The Bail System: An Economic Approach." In Gary S. Becker and William M. Landes, *Essays in the Economics of Crime and Punishment*. National Bureau of Economic Research. New York: Columbia University Press.

———. 1974b. "Legality and Reality: Some Evidence on Criminal Procedure." *Journal of Legal Studies* 3, 2: 287–337.

Leamer, Edward. 1978. *Specification Searches*. New York: Wiley.

Legal Aid Society of New York City. 1972. "Plaintiffs' Memorandum." New York Supreme Court, Appellate Division, First Department. New York: Legal Aid Society.

Levine, Daniel, Peter Stoloff, and Nancy Spruill. 1976. "Public Drug Treatment and Addict Crime." *Journal of Legal Studies* 5, 2: 435–62.

Lord, Frederick M. 1960. "Large-Sample Covariance Analysis When the Control Variable Is Fallible." *Journal of the American Statistical Association* 55, 290 (June): 307–21.

Lord, Frederick M., and Melvin R. Novick. 1968. *Statistical Theories of Mental Test Scores*. Reading, Mass: Addison-Wesley.

McFadden, Daniel. 1974. "Conditional Logit Analysis of Qualitative Choice Behavior." In Paul Zarembka, *Frontiers in Econometrics*. New York: Academic Press.

McKinlay, Sonja M. 1975. "The Effect of Bias on Estimators of Relative Risk for Pair-Matched and Stratified Samples." *Journal of the American Statistical Association* 70, 352 (December 1975): 859–64.

Madansky, Albert. 1959. "The Fitting of Straight Lines When Both Variables Are Subject to Error." *Journal of the American Statistical Association* 54: 173–205.

Maddala, G. S. 1983. *Limited-Dependent and Qualitative Variables in Econometrics*. Cambridge: Cambridge University Press.

Maddala, G. S., and L. F. Lee. 1976. "Recursive Models with Qualitative Endogenous Variables." *Annals of Economic and Social Measurement* 5, 4: 525–45.

Malinvaud, E. 1970. *Statistical Methods of Econometrics*. 2nd ed. Amsterdam: North Holland.

Manski, Charles F., and Steven R. Lerman. 1976. "The Estimation of Choice Probabilities from Choice Based Samples." Mimeographed. Pittsburgh: School of Urban and Public Affairs, Carnegie-Mellon University.

Morse, Wayne, and Ronald Beattie. 1932. "Survey of the Administration of Criminal Justice in Oregon, Report No. 1: Final Report on 1771 Felony Cases in Multnomah County." Oregon Law Review 11, 4 (supplement): 86–117.

Mosteller, Frederick, and Daniel P. Moynihan, eds. 1972. *On Equality of Educational Opportunity*. New York: Vintage.

Mosteller, Frederick, and John Tukey. 1977. *Data Analysis and Regression*. Reading, Massachusetts: Addison-Wesley.

Nagel, Stuart S., and Marian Neef. 1977. *The Legal Process*. Beverly Hills: Sage Publications.

Neyman, J. 1979. "Comment." *Journal of the American Statistical Association* 74, 365: 90–94.

Nicholson, Everard. 1970. *Success and Admission Criteria for Potentially Successful Risks*. Providence, Rhode Island: Brown University.

Nunnally, J. C. 1975. "The Study of Change in Evaluation Research: Principles Concerning Measurement, Experimental Design, and Analysis." In Elmer L. Struening and Marcia Guttentag, *Handbook of Evaluation Research*, vol. 1. Beverly Hills: Sage Publications.

Olson, R. J. 1980. "A Least Squares Correction for Selectivity Bias." *Econometrica* 48: 1815–20.

Pateman, Carole. 1970. *Participation and Democratic Theory*. Cambridge: Cambridge University Press.

Pechman, Joseph A., and P. Michael Timpane. 1975. *Work Incentives and Income Guarantees*. Washington, D. C.: Brookings Institution.

Rae, Douglas. 1971. *The Political Consequences of Electoral Laws*. New Haven: Yale University Press.

Ross, H. Laurence. 1975. "The Scandinavian Myth: The Effectiveness of Drinking-and-Driving Legislation in Sweden and Norway." *Journal of Legal Studies* 4, 2: 285–310.

Rothenberg, Thomas, and C. T. Leenders. 1964. "Efficient Estimation of Simultaneous Equation Systems." *Econometrica* 32: 57–76.

Sargan, J. D. 1958. "On the Estimation of Economic Relationships by Means of Instrumental Variables." *Econometrica* 26: 393–415.

Selltiz, Claire, Lawrence S. Wrightsman, and Stuart W. Cook. 1976. *Research Methods in Social Relations*. 3rd ed. New York: Holt, Rinehart, and Winston.

Sherwood, Clarence D., John N. Morris, and Sylvia Sherwood. 1975. "A Multivariate Nonrandomized Matching Technique for Studying the Impact of Social Interventions." In Elmer Struening and Marcia Guttentag, eds., *Handbook of Evaluation Research*, 1: 183–224. Beverly Hills: Sage Publications.

Smart, J. J., and Bernard Williams. 1973. *Utilitarianism: For and Against*. Cambridge: Cambridge University Press.

Struening, Elmer L., and Marcia Guttentag. 1975. *Handbook of Evaluation Research*, 2 vols. Beverly Hills, California: Sage Publications.

Sullivan, Walter. 1979. "Radical Surgery May Not Be Best for Breast Cancer." *Minneapolis Tribune*, January 30, p. 3B. (New York Times News Service)

Taylor, H. C., and J. T. Russell. 1939. "The Relationship of Validity Coefficients to the Practical Effectiveness of Tests in Selection: Discussion and Tables." *Journal of Applied Psychology* 23: 565–78.

Theil, Henri. 1958. *Economic Forecasts and Policy*. Amsterdam: North Holland.

Theil, Henri. 1971. *Principles of Econometrics*. New York: Wiley.

Thomas, Wayne H., Jr. 1976. *Bail Reform in America*. Berkeley: University of California Press.

Tufte, Edward. 1974. *Data Analysis for Politics and Policy*. Englewood Cliffs, New Jersey: Prentice-Hall.

Tukey, John W. 1977. *Exploratory Data Analysis*. Reading, Mass.: Addison-Wesley.

U.S. Department of Justice. 1965. *Proceedings of the National Conference on Bail and Criminal Justice*, Washington, D. C., May 27–29, 1964. Washington, D. C.: U.S. Department of Justice.

Warner, Dennis. 1976. "A Monte-Carlo Study of Limited Dependent Variable Estimation." In Stephen M. Goldfeld and Richard E. Quandt, *Studies in Nonlinear Estimation*. Cambridge, Mass.: Ballinger.

Welsh, J. Daniel, and Deborah Viets. 1976. *The Pretrial Offender in the District of Columbia*. Washington, D. C.: District of Columbia Bail Agency.

Werts, Charles E., and Robert Linn. 1970. "A General Linear Model for Studying Growth." *Psychological Bulletin* 73, 1: 17–22.

———. 1971. "Analyzing School Effects: Ancova with a Fallible Covariate." *Educational and Psychological Measurement* 31: 95–104.

White, Halbert. 1984. *Asymptotic Theory for Econometricians*. Orlando, Florida: Academic Press.

Index

Designer: Mark Ong
Compositor: Trigraph, Inc.
Text: 10/13 Palatino
Printer: Malloy Lithographing
Binder: Malloy Lithographing